Seeding the Positive Anthropocene

I0035509

Seeding the Positive Anthropocene

Philip McShane, James Duffy,
Robert Henman, and Terrance Quinn

edited by
James Duffy, Sean McNelis, and Terrance Quinn

AP

Axial Publishing
Vancouver

COPYRIGHT © 2022 James Duffy, Robert Henman, and Terrance Quinn
All rights reserved. No part of this publication may be reproduced, stored in a
retrieval system, or transmitted, in any form or by any means, photocopying,
electronic, mechanical, recording, or otherwise, without prior permission of the
copyright holders.

Axial Publishing
2-675 Victoria Drive
Vancouver, British Columbia
V5L 4E3 Canada
www.axialpublishing.com

ISBN   978-1-988457-10-9

Text layout and cover:
James Duffy

Seeding the Positive Anthropocene

Philip McShane, James Duffy,
Robert Henman, and Terrance Quinn

edited by
James Duffy, Sean McNelis, and Terrance Quinn

AP

Axial Publishing
Vancouver

COPYRIGHT © 2022 James Duffy, Robert Henman, and Terrance Quinn
All rights reserved. No part of this publication may be reproduced, stored in a retrieval system, or transmitted, in any form or by any means, photocopying, electronic, mechanical, recording, or otherwise, without prior permission of the copyright holders.

Axial Publishing
2-675 Victoria Drive
Vancouver, British Columbia
V5L 4E3 Canada
www.axialpublishing.com

ISBN   978-1-988457-10-9

Text layout and cover:
James Duffy

"What, **then**, is it like to live in the real world as starling scientific parts of the genetics of murmuration?"

–Philip McShane (1932–2020), *Questing* 2020C "The First Mansions," April 2020[*]

[*]The Questing essays are available at: http://www.philipmcshane.org/questing2020

Table of Contents

Preface..i

I. Stratification of the Earth System
Anthropocene or anthropocene?..3
Naming the First Peaceful Coexistence Colloquium.......................................7
Humans Impacting Earth Systems... 11

II. What Is an Opener
For Openers, Baby It's Cold Outside.. 21
The Noosphere.. 25
Thinking About New Ways of Thinking... 29
For Openers, What's Going On... 33

III. Beginning Economics
Economics in the Anthropocene: Blue, Green, and Other Colors 41
Strange Business in the Anthropocene 43
Toni Ruuska's Capital Ideas 47

IV. Shifting Probabilities
It's Getting Better and Better, Worse and Worse 55
Managings of History: Governings in the Positive Anthropocene................. 59
Down and Out on Planet Earth 65

V. Ant Essays
Ant Hop 71
Ant Hope.. 75
Ant Hopper 79
Ant Racks... 85

VI. The Need for Cyclic Thinking
Crecycling *Sustainability*... 89
Crecycling *Insight*... 97
The Masses and Sustainability 109

VII. Structuring the Reach Towards the Future
Structuring the Reach Towards the Future...115

Afterword..139
Appendix A: Abstract for "Structuring the Reach Towards the Future" 159
Appendix B: The Positive Anthropocene Conference...................................163
Index...167
About the Authors...171

Preface

James Duffy

Oh, mercy, mercy me
Ah, things ain't what they used to be, no, no, no
Radiation underground and in the sky
Animals and birds who live nearby are dying[1]

In recent decades, myriads of people living in various parts of the world have witnessed unprecedented flooding, ravaging bushfires, oxygen-poor 'dead zones' along coastlines, and a record-setting number of hurricanes. Unprecedented changes to the Earth's biosphere and its local ecosystems have landed us in a dire situation of *climigration* due to flooding, unbearable pollen seasons, and frequent warnings of risks due to UV radiation exposure. Human activities such as deforestation, burning fossil fuels, and manufacturing chemical compounds have been damaging the environment, and Earth System scientists claim we have crossed a boundary not just with respect to climate change, but also with respect to biodiversity loss, land conversion, and fertilizer use.[2] In these four areas we are now on the other side of the threshold of stability. There is something of a consensus that 'things ain't what they used to be,' and the window of opportunity for mending our ways is shrinking.

What do we do now?

The question is ambiguous. Who are we? We are quite a diverse crowd, spread out geographically and historically, as well as by age, temperament, formation, gender, experience, and creed. Currently the world population is approximately 7.9 billion people. What are we doing? Too many things to list, and in any case the list would have to be interpreted, storied, and evaluated.[3] What

---

[1] Marvin Gaye, *Mercy Mercy Me*, What's Going On? (Motown Records, 1971). Marvin Gaye wrote this song about the environment. It is the second single from his 1971 album, and it won a Grammy Award in 2002.

[2] Johan Rockström and Mattias Klum, *Big World, Small Planet* (New Haven: Yale University Press, 2015).

[3] These three tasks are represented by the letters I, H, and Di in the staircase diagram in McShane's Helsinki essay "Structuring the Reach Towards the Future" on page 107 below. See also "interpretation, interpreters"; "story, story checkers"; and "dialectic,

i

have we, an estimated 117 billion members of *Homo sapiens sapiens*, been doing *lately*?[4] What are we not doing that might be worth doing, "not merely to survive, but to thrive; and to do so with some passion, some compassion, some humor, and some style"?[5] Given the dimensions of the predicament we are in, how might we move beyond well-intentioned moralizing to making policies, planning, and implementing plans that are Earth-friendly?

A growing concern about the Earth's future underlies a proposal to apply the term *Anthropocene* to the current geological epoch. In 2000, Paul Crutzen and Eugene Stoermer introduced the term, after Crutzen had grown tired of hearing the current geological epoch referred to as the *Holocene*.[6] Eight years later, the British geologist Jan Zalasiewicz, who chaired the Stratigraphy Commission of the Geological Society of London, proposed that *Anthropocene* be used to name a formal geological interval. Together with three colleagues, he wrote "The Anthropocene: a new epoch of geological time?"[7] which led to the creation of the Anthropocene Working Group (AWG) one year later (2009).[8]

Advocates of "deep history" encourage anthropologists, archaeologists, linguists, geneticists, and primatologists to include in their narratives about human beginnings "prehistory," a term that some claim is an arbitrary boundary. While you do not need to be a professional geologist or paleobiologist to appreciate that the emergence of human laughter and longing were relatively recent events in the

---

dialecticians" in the index and Henman's essay in chapter II, "Thinking About New Ways of Thinking."

[4] How you or I read the word *lately* is a personal matter having to do with what time it is. *Lately* might mean roughly the last 260,000 to 350,000 years of mating, singing and dancing, making art, tattooing and applying makeup. See also my comments about *nowadays* in note 12 Of "Humans Impacting Earth Systems" on p. 14 below.

[5] "Maya Angelou: In Her Own Words," *BBC News*, May 28, 2014, sec. US & Canada, https://www.bbc.com/news/world-us-canada-27610770.

[6] Paul J. Crutzen and Eugene F. Stoermer, "The 'Anthropocene,'" *Global Change Newsletter* 41 (May 2000), 17–18.

[7] Jan Zalasiewicz et al., "The Anthropocene: A New Epoch of Geological Time?," *Philosophical Transactions of the Royal Society A: Mathematical, Physical and Engineering Sciences* 369, no. 1938 (March 13, 2011), 835–41, https://doi.org/10.1098/rsta.2010.0339.

[8] Crutzen and Stoermer proposed the latter half of the 18th century as the beginning of the Anthropocene, noting that the date coincides with Watt's invention of the steam engine. Geologists now believe that the significant indicator is plutonium from nuclear weapons testing. This moves the start date to the mid-20th century. See also note 20 below.

story of the Earth,[9] it is no laughing matter that during the last 70 years "it's gotten worse and worse, faster and faster."[10] Nor is it a laughing matter that the window of opportunity to mend our ways is closing.

Individuals and groups who are not geologists are also concerned about human behaviour and the Earth's future. People as different as Greta Thunberg, Pope Francis, and David Attenborough have been imploring us to reflect upon how future generations will survive. These and other individuals may or may not be interested in the efforts to formalize the Global Boundary Stratotype and Section Point (GSSP, or 'golden spike') for the Anthropocene. Likewise, those involved in the search may or may not be interested in what Greta Thunberg, Pope Francis, and David Attenborough say or write.

The "we" who are the four authors of the collection of essays in this book are part of a growing number of people concerned about what individuals and groups might or should do to mend our ways and live more responsibly so that our great-grandchildren find the planet in a better condition than we currently find it. We are not geologists, paleobiologists, or Canadian film makers,[11] but we are

---

[9] The origins of human laughter can be traced back 10 to 16 million years, while organic tools found in Ethiopia that might have symbolic meaning and reveal acts of reverence are approximately 1.4 million years old. Rounding off the age of the Earth to 4.5 billion years, cooperative and competitive behaviour encouraged by tickling has existed for approximately .003% of Earth story, while conscious longing has existed for approximately .0003% of Earth story. Marina Davila Ross, Michael J Owren, and Elke Zimmermann, "Reconstructing the Evolution of Laughter in Great Apes and Humans," *Current Biology* 19, no. 13 (July 14, 2009), 1106–11, https://doi.org/10.1016/j.cub.2009.05.028. Ran Barkai, "Lower Paleolithic Bone Handaxes and Chopsticks: Tools and Symbols?," *Proceedings of the National Academy of Sciences* 117, no. 49 (December 8, 2020), 30892–93, https://doi.org/10.1073/pnas.2016482117.

[10] Todd LeVasseur, "It's Getting Better and Better, Worse and Worse, Faster and Faster," Pasi Heikkurinen, ed., *Sustainability and Peaceful Coexistence for the Anthropocene* (Abingdon, Oxon: Routledge, 2017), 87–100. Hereafter *Sustainability*. Philip McShane reviews this essay on pages 55–58 below.

[11] Nicholas de Pencier, Jennifer Baichwal, and Edward Burtynsky directed the documentary "ANTHROPOCENE: The Human Epoch," a "meditation on humanity's massive reengineering of the planet" (https://theanthropocene.org/film). The film was released on September 13, 2018, in Toronto.

Philip McShane (1932–2020) was trained in mathematical physics. He wrote extensively in diverse areas, including evolutionary theory, linguistics, economics, and methodology. Terrance Quinn has published in pure and applied mathematics, science

interested in what they say, write, and produce. We are also interested in the contributions of others who share our concern. Finally, our interests include how all those concerned might proceed to ask and answer questions efficiently.

In the late fall of 2018, we took a specific interest in reading and reviewing the essays published in *Sustainability and Peaceful Coexistence for the Anthropocene*.[12] Our focus on this co-authored book was strategically limited, as we were intent on generating discussion and engaging the authors in dialogue.

At the time of writing the essays in this collection, two conferences were planned—*The 3ʳᵈ Peaceful Existence Colloquium* in Helsinki, Finland, June 13–14, 2019, and *The Positive Anthropocene* at the University British Columbia, Vancouver, Canada, July 8–12, 2019.[13] Emails were exchanged, McShane submitted an abstract for the June conference in Helsinki,[14] and soon after the organizers of that conference invited him to participate in the colloquium. The essay "Structuring the Reach toward the Future" is his contribution to the conference in Helsinki. McShane provides an image of the Anthropocene stretching thousands of years before 1950 and after 2050.[15]

---

education, and philosophy of science. In recent years, he has been doing work in economics and has published on ecological economics. Robert Henman did his graduate work in philosophy of education and has published articles in psychotherapy, meta-neuroscience, and ethics. I did graduate work in philosophy and have published on foundations in probability theory, economics, and the ethics of collaboration. More extensive biographies are provided on the last pages of this book.

[12] *Sustainability* is divided into four parts, and each of the four of us read and reacted to the essays in one of those parts: (I) Concepts, Causes, and Consequences (reviewed by Duffy); (II) Capitalism and Neoliberal Governmentality (reviewed by Quinn); (III) Thinking and the Non-Human World (reviewed by McShane); (IV) Post-Growth Societies and Organizations (reviewed by Henman). Our respective review essays are found in parts I–IV of this book.

[13] See Appendix B on page 163.

[14] See Appendix A on pages 159–161.

[15] The "central problem" of McShane's intervention in Helsinki is discussed in note 3 on p. 116. See also his description of axial humanity on 134–136. In note 12 of "Humans Impacting Earth Systems" (p. 14), I comment on the word *nowadays*. A stretching of the imagination is also intimated in "Anthropocene or anthropocene?" "The Noosphere," "For Openers, What's Going On," "Economics in the Anthropocene: Blue, Green, and Other Colors," "Managings of History: Governings in the Positive Anthropocene," "Down and Out on Planet Earth," "Crecycling *Sustainability*," and "Crecycling *Insight*."

The authors of *Sustainability* approached the challenge of understanding sustainable development from various fields of study, including geosciences, economics, law, organizational studies, political theory, and philosophy. In the words of Pasi Heikkurinen, the editor of the collection of essays, "the ideas presented in this volume are to work as an early invitation to a transdisciplinary dialogue under the rubric of the Anthropocene."[16] He goes on to describe the contribution the various authors make in these words: "The contributors provide important signposts from diverse perspectives on how things can, and should, be transformed."[17]

The essays in this collection also provide important signposts on possible transformations. Most of the essays originally appeared in the end of 2018 and beginning of 2019 as part of a website *Openers of the Positive Anthropocene*. There already existed and still exist various Anthropocene websites among the approximately 1.93 billion websites in existence (as of March 2022), but they were not and are not coordinated. A minor reason for the lack of coordination is the simple fact that websites serve different purposes. Static websites might inform, disseminate technical reports, announce events, preserve materials (archive), and/or digitalize a portfolio. Dynamic websites typically host forum discussions, instruct visitors to do something such as subscribing to a newsletter or downloading a document using a call-to-action (CTA), market and sell a product or service, gather information from possible clients, and/or organize a movement. Many websites combine static and dynamic elements. In addition, it is common for individuals to have web sites on which they do many of the aforementioned and, in some cases, provide a CV and social media links, for example a Twitter feed.

The four authors of this collection of essays maintain that the major reason for the lack of coordination is the not-so-simple fact that there does not exist an operative framework for collaboration, although there is an increasing number of people who are interested in figuring out whether and how it is possible to manage all the moving parts—a multitude of questions, methods of inquiry, time scales, disciplines, and worldviews. The lack of coordination is one of the pain points of our efforts, all our efforts.

The pain point might be felt in asking oneself, myself, ourselves: Where is this book, this essay, this colloquium, this formal definition, or this student

---

[16] *Sustainability*, 4.
[17] *Sustainability*, 4.

protest[18] "going" in the next few years, decades, or centuries? To whom is it going? How is it going? If it gains traction, might it be a small but important step, something that transforms the Earth for the better? Might it shift probabilities of seeding an "effective and resolute intervention" in world process?[19] These are some the questions raised in the essays in this book.

* * *

For many writers of essays and books, the term *Anthropocene* designates a period that is negative. There might be disagreement about when the "bad times" began,[20] but the wreckage is undeniable, and humanity must somehow get past the Anthropocene and jumpstart another, later period.[21] So, for most scholars, adding the qualifier *negative* is redundant, as the Anthropocene is a negative predicament. "Our planetary system is affected by a magnitude of force as powerful as any naturally occurring global catastrophe, but one caused solely by a single species: us."[22] For others, adding a value judgement *negative* or *positive*, while questionable, is tolerable, as long as such an evaluation is not confused with the

---

[18] In the spring of 2019, students around the world skipped classes to protest what they believed are government failures to intervene against global warming.

[19] Bernard Lonergan discusses the possibility of a "resolute and effective intervention in the dialectic" (historical process) in Bernard Lonergan, *Phenomenology and Logic: The Boston College Lectures on Mathematical Logic and Existentialism*, ed. Philip McShane, Collected Works of Bernard Lonergan 18 (Toronto: University of Toronto Press, 2001). 305–307.

[20] Proposed beginnings of the "bad times" include: the advent of agriculture, colonialism and plantation agriculture, the rise of capitalism, the Industrial Revolution, and the Atomic Age.

[21] See, for example, A. Fremaux, *After the Anthropocene: Green Republicanism in a Post-Capitalist World* (Cham, Switzerland: Palgrave MacMillan, 2019); A.Y. Glikson, *The Plutocene: Blueprints for a Post-Anthropocene Greenhouse Earth* (Cham, Switzerland: Palgrave MacMillan, 2017); E. Priyadharshini, *Pedagogies for the Post-Anthropocene: Lessons from Apocalypse, Revolution & Utopia*, Cultural Studies and Transdisciplinarity in Education, 14 (Singapore: Springer 2021); P. Heikkurinen, T. Ruuska, O. Rantala, and A. Valtonen, *After the Anthropocene: Time and Mobility* (Basel, Switzerland: MDPI, 2020); L. Young (ed.), *Machine Landscapes: Architectures of the Post Anthropocene*, (Oxford: John Wiley & Sons, 2019).

[22] Edward Burtynsky, "Artist's Statement" posted on "Photographs: Anthropocene," https://www.edwardburtynsky.com/projects/photographs/anthropocene.

value-free search for a golden spike or the practice of doing geological science or Earth science.

One way to reduce possible ambiguity would be to establish a convention: use capital "A" *Anthropocene* to designate a geological epoch—assuming it is eventually approved by the ICS and later by the International Union of Geological Sciences—and small "a" *anthropocene* to designate other meanings. [23]

Or should it be the other way around? Or …?

In this collection of essays, we distinguish between a *positive* and *negative Anthropocene*. While this may seem to be something of an anomaly, caring for the globe by removing *anthropos* from the scene would be a strange, not to mention unethical way of expressing care. A hermeneutics of suspicion about what has been and is going on, or not, might help diagnose the dire situation, but proposing a remedy requires a hermeneutics of recovery.[24] The dire situation complicates a possible recovery insofar as it makes it exceedingly difficult to fantasize grouped groupings of *anthropos* efficiently and beautifully flying together,[25] collaborating to "liberate many entirely and all increasingly to the field of cultural activities."[26] As

---

[23] See also "Anthropocene or anthropocene?" below, 3–5. In the Afterword I comment at greater length on well-intentioned efforts to manage meanings.

[24] The distinction between the two hermeneutics is found in the writings of Paul Ricoeur. A basic presentation of the distinction is given by Don Ihde, *Hermeneutic Phenomenology: The Philosophy of Paul Ricoeur* (Evanston, IL: Northwestern University Press, 1971), 140–143.

[25] On the front cover of this book, there is an image of starling murmuration, an amazing phenomenon that involves psychic adaptation in the flock. Each starling adjusts speed and velocity to that of its neighbors. See Andrea Cavagna et al., "Marginal Speed Confinement Resolves the Conflict between Correlation and Control in Collective Behaviour," *Nature Communications* 13, no. 1 (May 10, 2022), 2315, https://doi.org/10.1038/s41467-022-29883-4. This article focuses on statistical field theory. A more complete explanation would draw on and in genetic method to investigate, for example, the phrases "the essence of control, a fundamental problem of natural behavior" and "when the group is under perturbation." Ibid, 2. "What is history to tell us about the murmuration of starlings, a biodynamics that baffles present sciences? What is a later zoology to tell us about the subtle interbirding neurodynamics?" Philip McShane, "Questing2020C: The First Mansions," 2020, 4, http://www.philipmcshane.org/questing2020.

[26] Bernard Lonergan, *For a New Political Economy*, ed. Philip McShane, vol. 21, Collected Works of Bernard Lonergan (Toronto: Toronto University Press, 1998), 20. See further Quinn's essays in chapter III, "Beginning Economics" (41–52) and "Part Two: Remembrance of Times Past and Future" of McShane's essay "Structuring the

intimated in various essays in this collection, a bit of clarity regarding "what's what" and "what's what might be" would help seed the *positive Anthropocene*. Such luminosity would help coordinate the various efforts of data collectors, interpreters, story makers, evaluators and criticizers, policy makers, city and country planners, executors of plans, bloggers and journalists, science fiction writers, film makers, and other artists, and the millions of teachers and preachers around the globe trying to communicate pragmatic truths and hoping to implement timely precepts.

Reach Towards the Future" (127–133). See also Philip McShane, *Economics for Everyone: Das Jus Kapital* (Vancouver: Axial Publishing, 2017); Terrance Quinn, "Anatomy of Economic Activity," *American Review of Political Economy* 13, no. 1 (December 31, 2018), https://doi.org/10.38024/arpe.157; and James Duffy, "Minding the Economy of Campo Real," *Divyadaan: Journal of Philosophy and Education* 29, no. 1 (2018), 1–24.

I. Stratification of the Earth System

Anthropocene or anthropocene?

Philip McShane

There is an issue that needs attention by those who wish to arrive at an operative meaningful perspective on what is meant by the *positive Anthropocene*. I am using here, as a definite convention, the upper case "A". That convention, we should note, is not a settled convention, and there are those espousing a geological perspective who consider other perspectives on the name merit only a lower case *anthropocene* as a title for their concern.

This opens up a tortuous debate that is in fact beyond the present discussions of geological or other stratifications of evolution's path. Let me quote the beginning of section 2.8.1 of a recent discussion of the problem.

> The significance of the ICS International Chronostratigraphic Chart is that it provides an unambiguous definition of the geological column and provides a common language that scientists can use consistently. This emphasizes the importance of clearly defining what (upper-case) "Anthropocene" means in a strict, geological (and Earth System) sense and contrasting that with more general usages.[1]

The more subtle issue regards the meaning of *scientist* assumed in that paragraph. It is a meaning rooted in Aristotle's shrinkage of human understanding's explanatory efforts to the preliminary effort that is familiar from early geometry. "Throughout the whole of his works we find Aristotle taking the view that all other sciences than mathematics have the name of science only by courtesy, since they are occupied with matters in which contingency plays a part."[2] Sciences developed over the millennia under this restrictive cloud, so that, for instance, physics, dealing inevitably with contingencies, nonetheless became a respected science and anyone who pushed Aristotle still further could not be taken seriously: might one, even now, consider physics and chemistry as preliminary parts of engineering? Nor does the fact that one cannot do physics without engineering seem to bother the serious advocates of what may be called *pure physics*. And might I note that physics does not go forward without people to pose questions?

---

[1] Jan Zalasiewicz et al., "Making the Case for a Formal Anthropocene Epoch: An Analysis of Ongoing Critiques," *Newsletters on Stratigraphy* 50, no. 2 (April 2017), at p. 219, https://doi.org/10.1127/nos/2017/0385.

[2] W. D. Ross, ed., *Aristotle's Prior and Posterior Analytics* (Oxford University Press, 1949), 14. See also pp. 51ff. and the Aristotelian *Magna Moralia*, 1183b 9–18.

What, then, one might ask, is geology? How does a word that names people get into its domain? A little later in the section, mentioned already, "Making the case for a formal Anthropocene Epoch: an analysis of ongoing critiques," there is the remark, "the stratigraphic Anthropocene is founded on substantial changes to the Earth System that are reflected by an array of stratigraphic signatures."[3] What are the substantial changes to the Earth System, and how are they peopled and signed? The question I pose is rhetorical. I merely wish to draw attention to deeper puzzles regarding stratifications. That being neatly presented as something for a future fuller science of the Earth System, I venture to dodge the entire issue by placing the stage I call the *positive Anthropocene* as a stage of geology—like some, I would be content to take the date 1945 as a starting date, when some people were puzzled over the Japanese stratifications that occurred. What, then, is the positive Anthropocene? It is the period that began when humanity's stratification of the Earth System became tinged with flickers of human concern. It will rise to some maturity when such flickers have blossomed into a statistically effective science, a science, of course, of fulsome engineering.

I could go much further into this and into the objective "substantial changes to the Earth System," indeed go much further back to apes with a strange itch for habitat-shifting, but I am content to leave the naming business at that with a few little nudges. I would claim that the positive Anthropocene is preceded by a negative period, with two identifiable sub-periods. There is the early period when the substantial change of the Earth System that is the emergence of humanity was unaffected by humanity's attention to its own dynamic. There is the later period— perhaps associate its beginnings with Jaspers axial start[4] but extending it beyond our own times—when attention to that dynamic was an evolutionary sport that advances shabbily and distortedly: thus meriting the title *truncated negative Anthropocene.*

I would note finally, however, that the truncated negative Anthropocene is the scene of present discussions of the Anthropocene, be it lower or upper case.

---

[3] Zalasiewicz et al., "Making the Case for a Formal Anthropocene Epoch: An Analysis of Ongoing Critiques," 219–220.

[4] Karl Jaspers, *The Origin and Goal of History* (London: Rutledge and Kegan Paul, 1953). Jaspers places a basic axis of history in the period between 800 and 200 B.C., when humans reached significant differentiation in Greece, Persia, Israel, India and China. The end-date was disputed by both Toynbee and Voegelin. There emerges the view that the differentiation is still in its infancy and that its maturity is to be the seeding of the positive Anthropocene.

And it seems best now to begin taking a shot at the rescue of stratifications: we really cannot fiddle with names while homes burn.

Naming the First Peaceful Coexistence Colloquium

James Duffy

In the first chapter of *Sustainability and Peaceful Coexistence for the Anthropocene*, Pasi Heikkurinen recounts the story of finding a fitting name for the First Peaceful Coexistence Colloquium, which was a three-day gathering at the University for Peace in Costa Rica, April 27–29, 2015.[1] The theme of this colloquium was "Genders, Natures, and Technologies in the Anthropocene." Four of the ten chapters of *Sustainability* were presented at the conference, while the other six chapters were invited contributions.[2]

As chair of the conference responsible for putting together a call for papers, Heikkurinen had to come up with a name. The mission statement of the University for Peace caught his attention:

> … an international institution of higher education for peace with the aim of promoting among all human beings the spirit of understanding, tolerance and peaceful coexistence, to stimulate cooperation among peoples and to help lessen obstacles and threats to world peace and progress …

The phrase *peaceful coexistence* would eventually find its way into the title of the colloquium, while the word *progress* would not. Heikkurinen writes about the spin that is often times given, knowingly or not, to this problematic eight-letter word:

> Does it really refer to increasing the affluence and rights of humans at the expense of non-humans? Is progress here to signify the same kind of quantitative progress that the planet and its inhabitants have witnessed since the Industrial Revolution, that is: the development of more advanced means to transform the non-human-made world for the economic and political gain of the capitalist class?[3]

---

[1] A second Colloquium "Reimagining Ethics and Politics of Space for the Anthropocene" took place in the Finnish Lapland on June 6–9, 2017.

[2] The four chapters that were presented at the University of Peace are: 4. "Capitalism and the Absolute Contradiction in the Anthropocene" (Toni Ruuska); 5. "Managing the Environment: Neoliberal Governmentality in the Anthropocene" (Jessica C. Lawrence); 6. "'It's Getting Better and Better, Worse and Worse, Faster and Faster': The Human Animal in the Anthropocene" (Todd LeVasseur); and 7. "Scale, Noosphere Two, and the Anthropocene" (J. Mohorčich).

[3] *Sustainability*, 9–10.

Here it is worth mentioning three issues that intertwine in the Anthropocene, both as lived and as written about. One has to do with the limits of languages, which are always evolving. The second has to do with the possibility of multiple interpretations, some of which can go astray. The third has to do with the possibility of new meanings emerging through discoveries, breakthroughs, and differentiations.

An example of language evolving is the word "hacker" as it is used nowadays. The original reference to someone benignly working on a technical problem arose at MIT in the 1950s.[4] Since then it has been twisted to mean a person who secretly and maliciously exploits a computer system to gain access to information ("backdooring"). Other examples are the neologisms *climigration* and *clirefugee*,[5] *misothery*, a term coined by Jim Mason in 2006, the category *hyperobject*, created by Tim Morton in 2013,[6] and the term *crecycling*.[7]

The second issue is that while twisting and turning and sometimes distorting and corrupting language may occur here or there, in a handful of individuals, it can also occur on a larger scale. This was a concern of the Danish philosopher, theologian, poet, social critic and religious author Søren Kierkegaard (1813–1855), who asked himself if he truly was a *Christian*. A word, expression, or creed can be repeated by an individual or group with a meaning that might be true to the original meaning of an originating community, perhaps even better than the original meaning. But it can go the other way, in which case the meaning of a word, expression, or creed is distorted or corrupted by individuals or groups, some of them perhaps with good intentions. This was the concern of Heikkurinen regarding whether or not to include the "heavily loaded and bloody word *progress*" in the title of the colloquium. He decided not to.

A term that has also been debated, twisted, turned, and interpreted differently by different parties, one that Heikkurinen decided to include in the title of the

---

[4] Ben Yagoda, "A Short History of 'Hack,'" *The New Yorker*, March 6, 2014, http://www.newyorker.com/tech/elements/a-short-history-of-hack.

[5] Tarja Ketola uses these terms in *Sustainability*, chapter 3 "Climate Change Immigrants or Refugees of the Anthropocene—Adapting to or Denying Climate Change?"

[6] LeVasseur uses *misothery* a half dozen times and *hyperobject* thirty-eight times in in *Sustainability*, chapter 6 "'It's getting better and better, worse and worse, faster and faster': the human animal in the Anthropocene."

[7] McShane introduces this term in the second paragraph of "Crecycling *Sustainability*," 89.

conference, is the term *Anthropocene*. As he notes, those using the term tend to focus on "the imbalance between human and non-humans" but by and large remain silent on "the intra-species difference, that is: who *within* the human species is causing the destruction?"[8] Some think that the issue can be resolved by reserving uppercase *Anthropocene* to refer to the current phase of geological history as studied by the geological community, while reserving lowercase *anthropocene* to refer to changes in human culture, something similar to using the term *Renaissance* to indicate complex changes in human culture without considering geology or stratigraphic indicators.[9] Heikkurinen's reason to include *Anthropocene* in the conference title was to acknowledge and promote the complementarity of two types of analyses—those focusing on noteworthy changes in the Earth and those focusing on who, how, and why the human species is responsible for such changes. For example, the authors of chapter 2, "The Anthropocene: A Geological Perspective" implement the former focus, while the author of chapter 4, "Capitalism and the Absolute Contradiction in the Anthropocene" tends toward the latter focus.

---

[8] *Sustainability*, 11.

[9] See Zalasiewicz et al., "Making the Case for a Formal Anthropocene Epoch: An Analysis of Ongoing Critiques," 219–221.

Humans Impacting Earth Systems

James Duffy

The second chapter of *Sustainability and Peaceful Coexistence for the Anthropocene,* "The Anthropocene: a geological perspective," was co-authored by Mark Williams, Jan Zalasiewicz and Colin Waters, who are members of the Anthropocene Working Group (AWG).[1] Together they present a summary perspective on the ongoing search for distinctive stratigraphical evidence that may signal a new stage in the evolution of the Earth. The AWG includes specialists from different areas, including historians, archaeologists, social and environmental scientists. That being said, and as noted in the title, the perspective presented in this chapter is geological and highly specialized. The authors review the geological impact of humans on the Earth looking for a definite boundary in the stratigraphic record in the context of overall Earth time, which is more than 4.5 billion years. If the ICS one day finds the evidence compelling, then the term *Anthropocene* will become a new addition to the Geological time scale. On the other hand, "if it is not formalized the term will nevertheless likely still carry considerable currency across many disciplines as a way of discussing and debating the impact of humans on the Earth System."[2]

What constitutes significant impact? In geological terms, if a giant asteroid were to hit the Earth tomorrow, it would be of significant impact, leaving behind a widely traceable stratigraphic signal—assuming that whatters either survive the worldwide climate disruption or emerge from the wreckage to trace the signals. A similar pragmatic approach seeks a boundary event involving humans that would indicate a distinctive stratigraphical signal. So the search is basically for a

---

[1] The AWG was established by the Subcommission on Quaternary Stratigraphy, which in turn is a component of the International Commission on Stratigraphy (ICS). "The International Commission on Stratigraphy is the largest and oldest constituent scientific body in the International Union of Geological Sciences (IUGS). Its primary objective is to precisely define global units (systems, series, and stages) of the International Chronostratigraphic Chart that, in turn, are the basis for the units (periods, epochs, and age) of the International Geologic Time Scale; thus setting global standards for the fundamental scale for expressing the history of the Earth." "International Commission on Stratigraphy," accessed May 30, 2022, https://stratigraphy.org.

[2] *Sustainability*, 26.

"boundary in the stratigraphic record," one that is isochronous and "above which it is possible to recognize the profound influence of humans across the planet."[3]

What are some proposals for when the Anthropocene began? One is the beginning of the nineteenth century, at the dawn of the Industrial Revolution. This was a period of rapid population growth, urban development, and industrialization relying on an increasing use of fossil fuels. The 'problem' with identifying the early nineteenth century as the beginning of the Anthropocene is that industrialization is a process that took time to spread—indeed is still spreading in some parts—so "its accompanying stratigraphic signals are similarly time-transgressive, and so not ideally suited to best represent a geological time boundary."[4]

A second possible beginning of the Anthropocene is the mid-twentieth century. The term "Great Acceleration" was coined to describe a confluence of various indicators, for example, the rise in atmospheric carbon dioxide levels, a notable acceleration in agriculture (the Green Revolution) and in the use and spread of manufactured goods.[5] Two dozen graphs of Earth-system and socio-economic trends show a dotted line precisely at the year 1950.[6]

In conclusion, the authors note that there may be sufficient physical (e.g., sedimentary), chemical (e.g., nitrates), and biological (e.g., neobiota) evidence for claiming that the mid-twentieth century marks a turning point in the impact humans have had on the billion-year evolution of the Earth and all its critters.

And all its critters? Am I, are you, our ancestors, and descendants to be considered part of Earthly events and systems? The emergence of the human species (*Homo sapiens sapiens*) is quite recent—less than 200,000 years ago—given the geological record under investigation, one that is over 4.5 billion years. There is evidence in technofossils of tools made over 3 million years ago. The evolution of agriculture some 10,000 years ago changed not only the lay of the land, but "provided the environment in which human specialist activities unrelated to food production could evolve."[7]

---

[3] Mark Williams, Jan Zalasiewicz, and Colin Waters, "The Anthropocene: A Geological Perspective," *Sustainability*, 19.

[4] Williams, Zalasiewicz, and Waters, "A Geological Perspective," 24.

[5] Will Steffen et al., "The Trajectory of the Anthropocene: The Great Acceleration," *The Anthropocene Review* 2, no. 1 (April 1, 2015), 81–98, https://doi.org/10.1177/2053019614564785.

[6] See the appendices "Socio-Economic Trends" and "Earth System Trends" on pages 16–17 below.

[7] Williams, Zalasiewicz, and Waters, "A Geological Perspective," 18.

Geologists seek stratigraphic record of the impact of human activity (e.g., the invention of technology, the intensification of agriculture, the increasing reliance on cars). They do not, however, consider what's going on in the accelerators, the innovators and shapers of culture. It would 'simply' not qualify as an instance of "practical application to geologists" to seek, for example, a stratigraphic signal of the impact of Picasso's artwork or Aristotle's opera omnia. That, you might say, is what sociologists, psycholinguists, philosophers, and historians do.

My View

What might I say about the search for stratigraphic records to identify a defining moment of change in significant human impact on Earth and all its critters? Like my colleague Terrance Quinn commenting on neoliberal governmentality, I have reason to suspect that geological investigations have emerged from within, and remain part of, the current ethos.[8] The ethos I have in mind is the fragmentation of disciplines, misguided talk about "pure science," and an unverifiable assumption that sound methodology requires those studying the biosphere to prescind from studying the bionoosphere of their curious critterself. What I find is a partial or total eclipse of integral whatting in those trying to make sense of the mess we are in. This is a geobiohistorical indication that we are indeed living in the negative Anthropocene.

Integral whatting asks "mightn't it?" even about the past, doesn't it?[9]

What might happen in the next four thousand or four million years in the bionoosphere? We hardly have a clue, but fantasy is essential for survival. There is evidence of whatting that is a fundamental change and changer in the cosmos. "The emergence of humanity is the evolutionary achievement of sowing what among the cosmic molecules."[10]

A fine opportunity for those concerned about life on Earth in 2040 or 2050, is to sow the question "what is sane economics?" into the geobionoosphere. It is quite evident that the 1970 Nobel Laurette Paul Samuelson, together with Nobel winners Amartya Sen (1996), Joseph Stiglitz (2001), and Paul Krugman (2008)

---

[8] See "Our Views and Hopes," 62–64.

[9] Consider how doing research on those working in a nineteenth-century chemistry search for a quinine treatment of malaria leads one spontaneously to wonder if the treatment mightn't have cured.

[10] See "The Noosphere" (26), the fifth paragraph of "Ant-Hop" (71), and the beginning of the Epilogue of "Structuring the Reach Towards the Future." (134)

missed basics insights that will be obvious to high schoolers in 100–150 years.[11] The lacuna has had, and continues to have, an impact on the Earth and all its critters.

Concern about high school education in 2120–2170 raises difficult questions about policies, plans, and communications reaching the masses, in this case high school administrators, teachers, and students. How might that happen? Geologists, like many scientists *nowadays*,[12] are not thinking about or encouraging the science fiction of a possible acceleration of 'human specialist activities unrelated to food production,' an acceleration that will "add aggregate leisure" and "liberate many entirely and all increasingly to the field of cultural activities."[13] An addition of leisure that liberates humans to become a bit more human—sounds like progress to me—and slowly yields an acceleration of and in and by the accelerator is 'simply' not part of current pragmatics. And it is extremely challenging to fantasize how a new pragmatics might yield fruit in empirically rich economics textbooks and non-truncated economics teachers in your high schools and mine in Mexico, say in 100 years.

As it is, the proposal to add the term *Anthropocene* to the Geological Time scale is at the "level of the times,"[14] the negative Anthropocene in which we live and move and think and breathe. One of the indications of this level is a blind spot regarding *how* the human species might effectively learn from history to plan.[15] Who among us agrees sufficiently with Arne Næss that "the complexity-

---

[11] This is the position I take in Duffy, "Minding the Economy of *Campo Real*." See also Quinn, "Anatomy of Economic Activity" and the Preface to McShane, *Economics for Everyone*, i–xiii.

[12] This is not an easy term to define, for it begs the question: "Where and when are we in the big picture, the story of the universe, approximately 13.7 billion years young?" By 'nowadays' I mean a time period of roughly seven thousand years: 4,000 years of history and 3,000 of future in which the wondering, whatting, planning two-legged, laughing animal is ever so slowly becoming luminous about wondering, whatting, and planning. A context is Philip McShane, "The Big Bang," in *A Brief History of Tongue: From Big Bang to Coloured Wholes* (Halifax, NS: Axial Press, 1998), 15–48.

[13] Lonergan, CWL 21. 20.

[14] "La altura de los tiempos" ("The level of the times") is a phrase used by the Spanish philosopher and essayist José Ortega y Gasset (1883–1955) in chapter 3 of *The Revolt of the Masses* (London: W. W. Norton & Co., 1932).

[15] See Tarja Ketola, "Climate Change Immigrants or Refugees of the Anthropocene – Adapting to or Denying Climate Change?," in *Sustainability*, 31–48.

not-complication principle favours division of labour, *not fragmentation of labour*"[16] to creatively fiddle with current academic trends of teaching, researching, and publishing?

The authors write that "the stratigraphic character of the Anthropocene may appear distinctly different from a far future perspective."[17] They have in mind changes in how stratigraphic signals of human impact on Earth Systems might be interpreted in the future. Insofar as they do not consider, nor consider the possibility of considering, how changes in the bionooosphere will impact the biosphere—for example by eliminating commerce and industry or transforming agriculture into a superchemistry[18]—their research is at the level of the times. Study of fossils and technofossils might indicate "progression in technological development from early stone, brick and ceramic technologies that culminate in the gigantic complex plastic, glass, metal, silicon and concrete structures of modern cities,"[19] but what if the study suffers from an eclipse of integral "mightn't it?"?

The challenge here and there and everywhere is for you and me to read the question marks ending the prior paragraph. There is a sense in which the negative Anthropocene has educated us out of our primitive pragmatic minds, making it difficult to fantasize a stage in the evolution of the Earth when humans find favor in dividing up labor and become geohistorical agents of progress.[20]

---

[16] Arne Næss, "The Shallow and the Deep, Long-Range Ecology Movement: A Summary," *Inquiry* 16, no. 1–4 (January 1, 1973), 97, https://doi.org/10.1080/00201747308601682. This principle of Næss is cited by McShane in "It's Getting Better and Better, Worse and Worse," 58 and in "Crecycling *Sustainability*," 89.

[17] Williams, Zalasiewicz, and Waters, "A Geological Perspective," 25.

[18] Lonergan, CWL 21, 20.

[19] Williams, Zalasiewicz, and Waters, "A Geological Perspective," 21.

[20] An image of divided-up agency is provided in the ten-step "staircase" diagram that McShane created and that appears with notes on p. 107 and without notes on p. 118 below.

Appendix A[21]

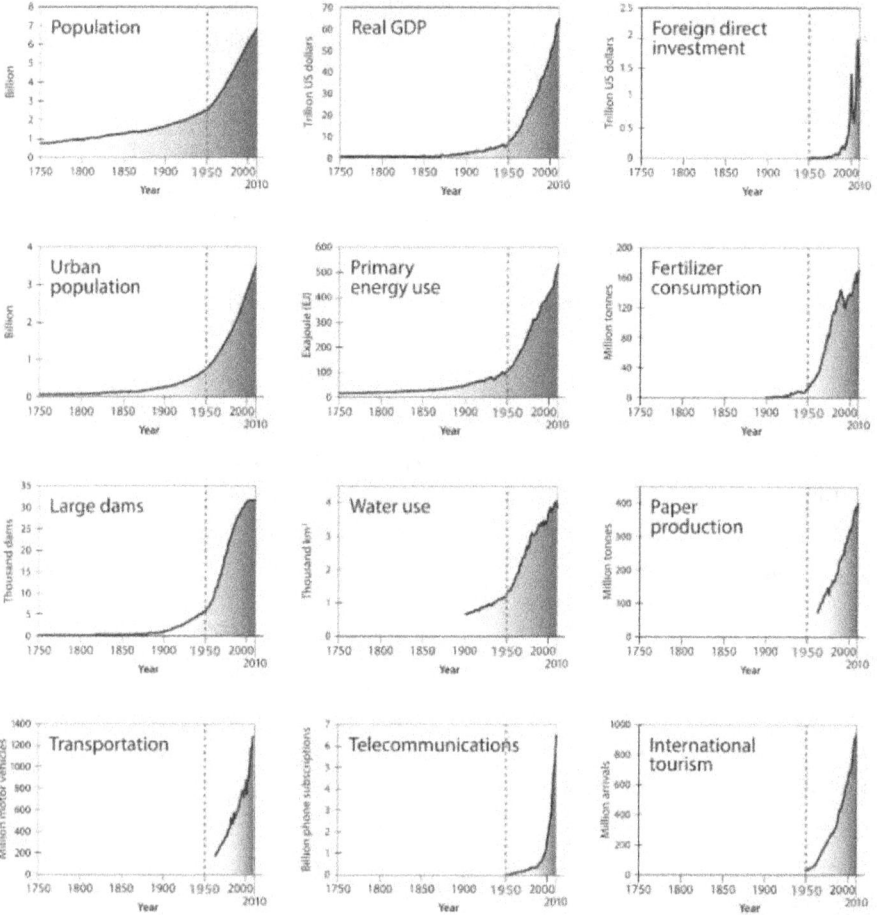

Socio-economic trends

Population · Real GDP · Foreign direct investment · Urban population · Primary energy use · Fertilizer consumption · Large dams · Water use · Paper production · Transportation · Telecommunications · International tourism

---

[21] Steffen et al., "The Trajectory of the Anthropocene," 4.

Appendix B[22]

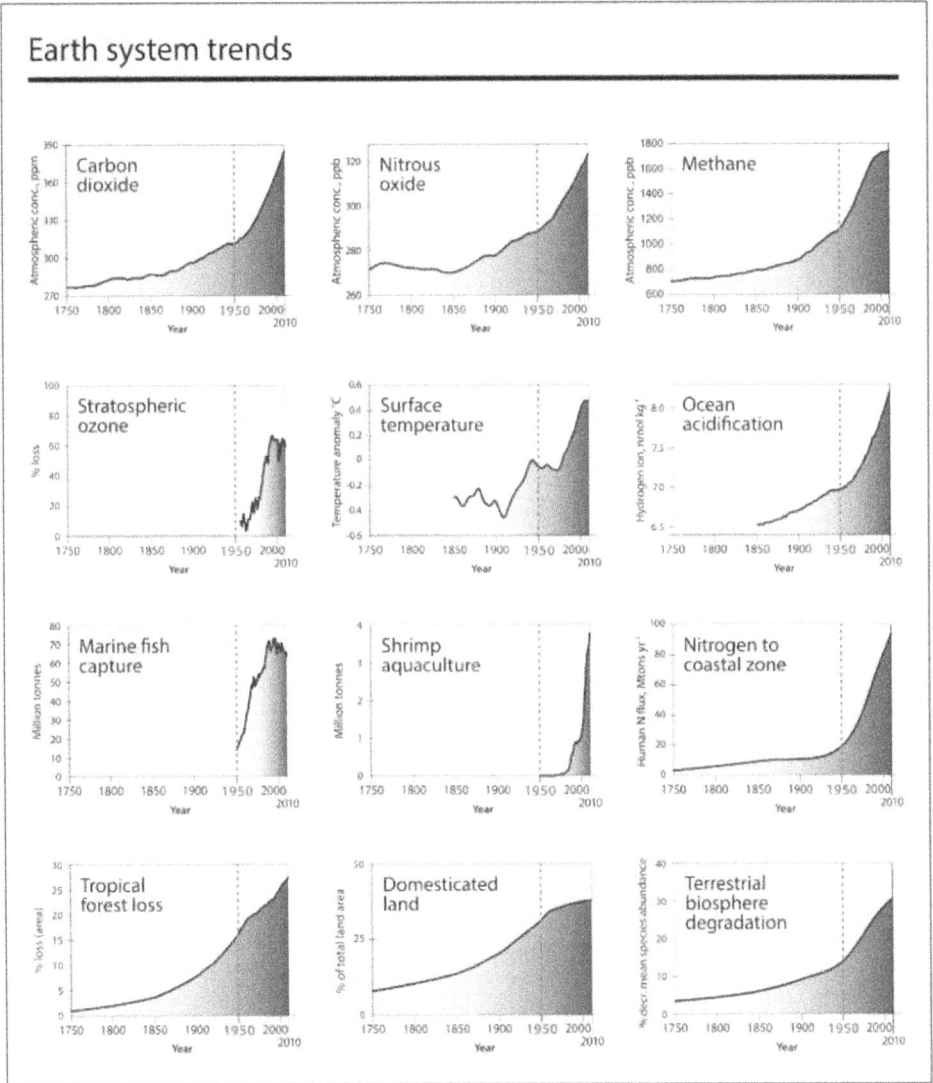

Earth system trends

[22] Steffen et al., "The Trajectory of the Anthropocene," 7.

II. What Is an Opener

For Openers, Baby It's Cold Outside

James Duffy

I really can't stay (Baby it's cold outside)
I gotta go way (Baby it's cold outside)[1]

For openers, here in Morelia, Michoacán it is cold and rainy, and the meteorologists say it will be this way for some days. A friend tells me that her mother told her that November would be colder and wetter than normal, and December will be more of the same. There is no indoor heating in these parts of Mexico, so the song lyrics need to be amended: Baby it's cold inside and outside.

Does global warming mean harsher winter weather? Do cold spells really call into question human-induced global warming?

Climate change has been in the news recently. The World Meteorological Organization says that 2018 is set to be fourth hottest year since records began. US President Donald Trump casts doubt on a report by his own government warning of devastating effects from climate change. He does not believe the report. Apparently, he gives himself more credit than he does to those scientists who differentiate between climate and weather.

The graph on the top of the next page of the global average surface air temperature has been used as the indicator of the state of the climate system.[2] It shows the warming of the climate system, including the strong rise from about 1970 to 2000.

---

[1] "Baby, It's Cold Outside," Frank Loesser, 1944.
[2] This is one of various graphs in Appendix B to "Humans Impacting Earth Systems," 17.

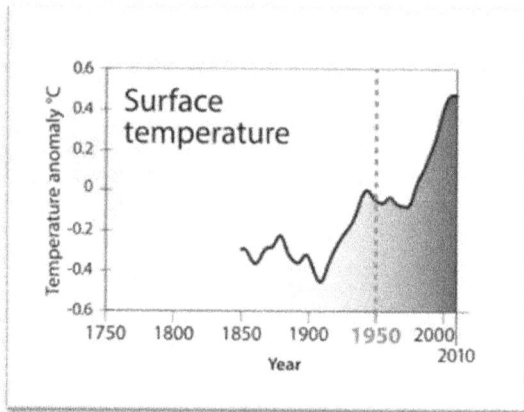

Earth-system and socio-economic post-1950 trends have been simulta-neously sweeping across both spheres of the Earth. Here I will not quibble over whether the two spheres are really one, nor will I weigh in on whether and when *Anthropocene* should be spelled with a capital "A" and whether and when *anthropocene* should be spelled with a small "a."[3] (Microsoft Word spell checker prefers capital "A.") Finally, neither will I attempt an evaluative interpretation of the various graphs. Explaining why the GDP[4] is a bogus indicator, for example, would sidetrack me from this opening reflection on openers. Instead, I simply highlight the number of things taken into consideration in these graphs—shrimp, nitrogen, fertilizer, water, methane, foreign investments, telecommunications, population. Wow! Other things might be considered, e.g., money spent on armaments or toilet paper consumption, but the graphs give an impression of the complexity of caring for the Earth.

For starters, something significant has happened and continues to happen, something that has put human survival at risk. Experts do not agree on a specific starting date of the new epoch—1610, early 1800s, 1950s, 1964—but there is a growing consensus that humanity is facing unprecedented challenges, a consensus that the clock is ticking and that we do not have a whole lot of time to get our act together before we find Earth uninhabitable. Among these challenges are mass extinctions of plant and animal species, polluted rivers and oceans, and alterations to entire ecosystems. Will Steffen, who heads Australia National University's

---

[3] See Zalasiewicz et al., "Making the Case for a Formal Anthropocene Epoch: An Analysis of Ongoing Critiques," 219–21.

[4] See "Real GDP" in Appendix A to "Humans Impacting Earth Systems," 16.

Climate Change Institute, writes of the name *Anthropocene*: "[It] will be another strong reminder to the general public that we are now having undeniable impacts on the environment at the scale of the planet as a whole, so much so that a new geological epoch has begun."[5]

For openers, we have inherited a mess, and it's not going to do any good to point the finger at dead American or European philosophers or at inept world leaders. So, what's to be done in and with the situation? How are we to assume responsibility? What are the challenges and opportunities in education?[6] Is there a way, a framework, a method for bringing various groups—stratigraphy experts and Earth System scientists, photographers and film directors, urban planners and ecological economists—into dialogue? Are the large challenges of changes in climate or ocean garbage toxicity twined with subtler challenges in educating Rita and Ralph that would support what's to be done by cultivating a new aesthetic, be it of dance or film or poetry?

In "For Openers, What's Going On," I muse about early education for openers, where "openers" are young, middle-aged, or old people, living in any hemisphere, concerned about their children and themselves surviving, not, then, just learning to live in the Anthropocene,[7] but searching for small steps to transform a negative epoch into a positive one.

---

[5] Cited by Joseph Stromberg, "What Is the Anthropocene and Are We in It?," Smithsonian Magazine, January 2013, https://www.smithsonianmag.com/science-nature/what-is-the-anthropocene-and-are-we-in-it-164801414.

[6] See Reinhold Leinfelder, "Assuming Responsibility for the Anthropocene: Challenges and Opportunities in Education," *Rachel Carson Center Perspectives* 3 (2013), 9–28.

[7] See Susan Laird, "Learning to Live in the Anthropocene: Our Children and Ourselves," *Studies in Philosophy and Education* 36 (2017), 265–82.

The Noosphere

Philip McShane

I begin by homing in on the problem zone that is the focus of our Anthropocene attention. I do that, in an introductory fashion, by linking our efforts to chapter 7 of the book, "Scale, Noosphere Two, and the Anthropocene," by J. Mohorčich. Let me move forward by quoting his beginning:

> Is it productive to think systematically and globally about thought the way it is productive to think globally about life or water? Terms like biosphere or hydrosphere connote a spherical, holistic understanding of the interrelation of ecosystems and hydrological systems. They coordinate ways of thinking about organisms and water at different scales. Though these spheres of understanding—lithosphere, magnetosphere, pedosphere, cryosphere and atmosphere—are familiar intellectual artefacts, the question of what sort of knowledge can be organized under the -sphere rubric remains open.[1]

Later in the article we find goodly suggestions about "loops" and "looping" as supplementing the "sphere" notion, and indeed, we, the group of four, hope to push forward this looping notion towards a fulsome contribution to the global effort of evolutionary progress to which the entire book points. But here it is a matter of intimating that there are fundamental issues of progress that need to be addressed if the efforts are to be increasingly successful in the genesis of peaceful progressive sphere-looping.

Mohorčich's article points to the less peaceful poises. The previous essay puts it well with a quotation from *Deep Green Resistance*: "One of the primary functions of government is to protect the rich sociopaths from the outrage of the rest of us."[2] We cannot expect the rich sociopaths to go quietly into the night or day. Indeed, McBay et al. draw attention to "Deep Ecological Warfare" with a manual titled *Deep Green Resistance*. But what, might we not ask, is this "drawing attention?"

So we arrive back, a simple loop in our little noosphere, at the beginning of Mohorčich's article. Pause—is this not a drawing of attention? —over the first sentence as—again I draw, or try to draw, attention: "Is it productive to think

---

[1] Heikkurinen, *Sustainability*, 101.

[2] Derrick Jensen, Lierre Keith, and Aric McBay, *Deep Green Resistance: Strategy to Save the Planet* (New York: Seven Stories Press, 2011), 388.

25

systematically and globally about thought the way it is productive to think globally about life or water?"

What do you think? Do you think, What? I have drawn in, on, out, your attention, have I not? And are you perhaps poised to answer, if only vaguely, "Yes," to the question posed about thinking systematically? What might it be like, to think systematically about thought? And think further: would this not point towards thinking systematically about all the thought that lurks behind and in this book?

Let me nudge further, leaning on your patience at this loopy looping round your noosphere. Is drawing attention not something like getting you to ask "What?" Again, think of that first sentence of Mohorčich's article that ends with a question mark. It is an Is-question: but there lurks a what-question in your attention to it, for otherwise you would not know what he is talking about. Not "know what"?

Patience please. "Know what" brings us to the kernel of our contribution to the struggle. Indeed, it points us towards a distinction that we make between the present negative Anthropocene and the positive Anthropocene, which is to come, a Capital Anthropocene where *Capital* and *Capitalism* take on, peacefully, radically new meanings.

Is "knowing" capital? This volume takes an implicit stand on it being indeed such. Should we not, then, know this capital, if it is the seed of evolution's hope? A few pages further on Mohorčich seems to be in clear agreement. "The noosphere is a name for the circulation of ideas, and the loops and networks they create as they attempt to understand themselves, each other and the conditions of their circulation."[3] "Attempting to understand themselves" is not easy to read sanely as referring to anything but the contributors of questions and ideas and loops and networks leading to what I call the *positive Anthropocene*. Are we back freshly at the appeal of Socrates to the Athenians, people already settled in destructiveness of the earth, to know themselves? I recall my own pointing in that direction as key to evolution's reorientation:

> The emergence of humanity is the evolutionary achievement of sowing what among the cosmic molecules. The sown what infests the clustered molecular patterns behind and above your eyes, between your ears,

---

[3] Heikkurinen, *Sustainability*, 112.

lifting areas—named by humans like Brocca and Wernicke—towards patterned noise-making that in English is marked by "so what?"[4]

The issue of eventual peace is a matter of finding our what's what, so that we are not blindly sowing what, as we have done right through the negative Anthropocene, however long we think it is. Our initial attempts, through this century, certainly will involve conflicts on all levels. But if we are to move beyond that messy and miserable and mean stage, we must find the basis of therapy for, e.g., the sociopaths, and that therapy requires that we loop around our whats and theirs in order to identify what went wrong in their ontic and phyletic evolution. Think Donald Trump. Think him through and true. Yes, I have surely caught your attention and indeed your agreement that such a thinking is virtually impossible at present. But pause again. There will be other trump cads, worse than Donald in the "misothery" (a term from the previous chapter 6, for total alienation) that is a sickness of our little globe in the cosmos.

I had best end these few rambling introductory words. My little essays on Ants[5] supplement this pointing to our needs and our entrapments. Do these words and those essays not, perhaps, help toward looping round Mohorčich concluding words?

> To consider the new noosphere is also to see it as a made object that subsists on physical conditions, not a mind floating frictionlessly above the world. This allows us to argue about what we want this made object to look like, how we want it to function, whose contributions we emphasize. The noosphere's growth is not, as in certain technologist-utopian accounts of a 'coming singularity', an inevitable and self-propelled drive to the exit of want.[†] It is partly self-propelled, yes, but its growth is implicated with consumption, strife, scale, pollution, identity, metabolism, peace, conflict, exploitation, communication, innovation and regression. To understand this is 'to shift from the Globe to the loops that slowly draw it', and to see these self-drawing ribbons at scale, unreeling in all directions.[††6]

---

[4] Philip McShane, *The Allure of the Compelling Genius of History: Teaching Young Humans Humanity and Hope* (Vancouver: Axial Publishing, 2015), 3.

[5] See part V "Ant Essays," 69–85.

[6] Heikkurinen, *Sustainability*, 113. The internal citations are to (†) Ray Kurzweil, *The Singularity Is Near: When Humans Transcend Biology* (New York: Viking Penguin, 2005) and to (††) Bruno Latour, "Facing Gaia: A New Inquiry Into Natural Religion" (Gifford Lectures, University of Edinburgh, 2013), 94, https://www.giffordlectures.org/lectures/facing-gaia-new-enquiry-natural-religion.

Thinking About New Ways of Thinking

Robert Henman

My contribution to this collection of essays is a focusing on education in Part IV of the book *Sustainability*. The focus raises the question: When we describe what we think we need as characteristics of the good life or what is needed to get us there, are we somehow putting the cart before the horse? As you and I read through chapters 8 to 10 of part IV of the text, *Sustainability*, we find that listing what is needed dominates much of the expression. The listing is the cart, and if you like you can think of it being pulled along. What does the pulling? What is the horse? Really, we somehow don't bother about identifying the horse. After all, we implicitly claim, is that not how we tend to live—envisaging our future with spontaneity—working out how to attain or achieve some revision of our lives? But in the present case the revision is seen to be quite beyond us, certainly when taken in terms of effectiveness.

Might I suggest that the spontaneity—not asking about the horse—is part of the Anthropocene in a prevailing negative state, subtly premised on a fairly static notion of history—business as usual forever. Thinking about what is needed cannot, oddly, get us beyond the static state. It stirs curiosity in readers and in Monday-morning quarterbacking but somehow does not foster an effective frontline challenging of either its own inadequacy in thinking or of the static state of a world spiraling down in settled resistance to change. Elaborating on what is needed is not heeded by the power brokers and seems to sow little in the searchers for change. Something more is obviously 'needed'. The vision of solution needs some new form of persuasion, some new piece of the listing that would wake up, not just the power brokers, (they may have to die off), but the listers.

Am I just adding confusion to your reading and mine, to the listers and the lists? Am I struggling still with the content of the cart and not really identifying the horse? Indeed, I am: and I add confusion by asking you to pause over the suggestion that the horse is in the cart.

Is this some daft Zen twist, or is it just muddy alchemical thinking? Let me here call on Bonnedhal. In section 3.1.4 of chapter 10 of *Sustainability*, Bonnedhal speaks of the transformation of our values in nature as commodities. He goes on, in section 3.1.5, to speak of the alchemical character of this thinking. On page 179 he writes that "The knowledge needed to make such decisions is unavailable." What form of knowledge would challenge alchemy? Bonnedhal speaks of such knowledge in his final statement of his chapter; "…we 'just' need to prove that we are morally and biologically apt to change."

So, I move to what may seem even crazier twists. What is this need to prove? Or might I not say, "Where is this need to prove?" And I invite you to leap away from "prove" to "approve" when I bring you back, horse-backed, to the odd claim, the horse is in the cart. Let me twist further by dropping the question mark that is normally at the back of the statement, thus writing "what is in the cart."

Here, I claim, is the seed of thinking about new ways of thinking about new ways. The battered and neglected piece of us that is wonder, that "what's?" proves its presence in listings but needs now open approval, indeed an approval that would turn its presence into a science, a science of the global future. The alchemy of this new chemistry of our survival is around: we have, in this little book, evidence of it in its listings and its talk of needs. But can we be tilted towards thinking seriously and effectively of this new chemistry, this new physics? Newton wondered about why apples fall and the moon does not. Einstein wondered how things could be as they are without absolute space or time. But what they both did, as did those between them in our human story, was to add to a successful listing of wonder-achievements, many of which battered human wonder in our industrious revolution. Neither Newton nor Einstein nor those between in that frenzy of wonder-destroying greed, recognized that wonder was fighting back in a deep alliance with nature—for wonder is nature at its best.

Various essays in this volume point to a split between humanity and nature, but the split is part of the alchemy of blinded wonder. The split is walking on the face of nature, of Gaia, of the molecular dynamism of wonder. Our challenge, yours and mine in this sentence that hangs over us, hangs in our neuromolecules, is to hear the hanging hanging round our lonelinesses and frustrations as we read the lists of evils and failed goods that *Sustainability* provides. But the provision, as I noted at the beginning, is trapped in a negativity of the Anthropocene, whose identification I have been weaving towards here, as have my colleagues with me. We are all trapped in a low-level culture of trivial living and listings. How are we to lift ourselves up out of that level, so that the future would inherit the wind that brings breathable air?

In McShane's essays we read of his image of the climb in scientific method as an Archimedean screw towards a crecycling (creative recycling) of our efforts. And to be sure, they are efforts and will remain so for some time. But that time can be shortened by involvement in a creative recycling of the division of labour that puts you and me inside the operations as operators. This insider perspective, of being at home in our wondering, sets us on a new path of thinking new ways about thinking new ways. As the Archimedean screw lifts water up the hill, so our wonderment, embellished in crecycling, lifts our efforts towards a higher probability of intervening in history and a more positive footprint by the human species.

Thinking about new ways of thinking about new ways is about education as being an educed cherishing of the wonderers we are. The educing can begin now, in and from the discomfort of this reading of your cart and heart. If we want to challenge the alchemy of the static notion of living and listing that consumes us, we will 'need' to begin wondering and wandering around the question, the horse in the cart, the cart that clamors for fresh travels in time. This is an old need that clings to our song and dance. Might we effectively notice that this old need clings to our listing of the characteristics of the Anthropocene, so that the negative darkness in our listings and our lists can break forward to a positive Anthropocene?

For Openers, What's Going On

James Duffy

We are openers. We open doors, windows, and boxes. In one week, some people will open presents. Stores, museums, and galleries open (and close), as do buildings and roads. They do not open by themselves, so perhaps it would be better to use the passive voice: stores, museums, buildings, and roads are opened by openers. Doors sometimes open automatically, but they do not design or produce themselves to function thusly.

One of the ways we open—perhaps unique to humans—is by posing questions. I have in mind the curiosity of a five-year-old boy, the son of a friend. When he visits, his what-poise is focused on having a good time, and he spontaneously asks: What are we going to do to have fun? Play hide 'n seek in the park? Race cars? Make paper airplanes for a flying competition? His father is usually focused on eating first, playing later. His spontaneous what-poise is more along the lines of "What do I say to my son to convince him that eating first and playing afterwards is the best option?"

It's safe to say that asking what-questions occurs in nearly all situations throughout the busy or even not-so-busy day. Getting practical chores such as cleaning and shopping done is permeated by what-questions. Likewise, whatting regularly occurs in the workplace, quickly or slowly, depending on the type of work. An air-traffic controller needs to be on their what-toes for the well-being of pilots and passenger alike. A lab researcher is also on their toes, but they probably have more time to deliberate, consent, choose, and perform. A chocolate or wine taster's whatting focuses on aesthetics—color, swirl, smell, taste, and savor. There are jobs were whatting is not so important, e.g., getting paid to sleep.

For many of us the best part of the day are moments of twofold liberation that occur, for example while watching a film or listening to music, when we are delivered from both biological purposiveness and from practical and occupational whatting. But even in a leisurely game of poker, where players have to decide if the cards in their hands are of sufficient value to open the betting, what-questions are rapidly self-posed: What might be a good opening bet? Why did she pause before raising the ante? What might she have in her hand? What does that look on his face mean?[1] What are the chances he is bluffing? What should I do next?

What goes on not just in five-year-olds in the park and poker players around a card table, but elsewho and elsewhere. It might very well be going on in you as you read along right now. What is the point of this essay? What is James getting at? Why does he use words that do not exist like *elsewho*?

In his essay "Ant Hop," McShane recalls Archimedes whatting to figure out how to get water to go uphill. Henman, in his essay "Down and Out on Planet Earth," cites Farah's Foreword and raises what I consider important questions for opening the positive Anthropocene: "Farah raises the point well in stating that: 'Modern societies should simply learn' and that seems to be the crux, we don't learn. Why is that? What keeps us from the type of learning required?"[2]

We should simply learn, but we don't. Why is that? When I ask university students what they learned in the prior semester, more often than not they do not have much to say. They memorized short and long names, some of them definitions, to pass exams, but normally they did not reach an understanding that passed into the texture of their mind and body. Some students intuit that naming is only a beginning, a starter, but it is not the same as the understanding that would allow them to teach a classmate asking questions that they themselves once asked and answered.

I am not saying that speaking about or listening to someone else speak about "four grey elephants" is an unintelligent activity. I am saying that reaching for the most up-to-date understanding of numbers (e.g., polynomial time algorithm to identify primes), colors (e.g., chromatics, color psychology), or animals (e.g., biodiversity and population dynamics, history of migration patterns) takes more time and much patience. I, for one, cannot speak about the biochemistry of African elephant mating habits. I would not know where to begin.

What if a young man or woman were interested in zoology or number theory or international development? What advice would I give him or her, for openers?

For Openers 1

Work on exercises that highlight the difference between using a name and understanding what the name means. This might be counter-cultural, depending

---

[1] At the world series of poker, normally one or two players are wearing dark sunglasses to hide their eyes and make it more difficult for other players to read them.

[2] See p. 66 below.

on one's context of teaching and learning. In areas such as physics there exists an ethos of patiently struggling to understand, and time set aside to experiment with inclined planes or swinging pendulums. In other areas, naming, sometimes sophisticated naming, overshadows understanding. The surrounding ethos might deceive you or me into believing that naming an emotion—*fear, surprise, anger*—is pretty close to understanding what the name means. The ethos might also suggest that understanding human emotion is easier than understanding the motion of an apple falling from a tree.

For Openers 2

Implement convenient diagrams, symbols, and heuristic devices to support whatting. These things are like topographic maps—they cannot replace the adventure, but they can provide orientation. Heuristics are symbols that guide the search for an unknown, some more convenient than others. For example, it is easier to take the square root of 1764 than MDCCLXIV. The notation dy/dx developed by Leibniz is more useful than $\dot{x}$.[3] The periodic table is a collection of convenient symbols that emerged in the late 18th century, was formally published in the second half of the 19th century, and since then has helped countless students and researchers to manage the meaning of *water, nitrous oxide, carbon dioxide,* and *methane* that appear in the socio-economic and Earth system trends.[4] Diagrams and other heuristic devices are tremendously helpful for cultivating a whatting ethos, a counterculture to the culture of naming and memorizing.[5]

For Openers 3

Pay attention, notice, take note of, and appreciate instances of what is represented in the diagram on the top of the next page.[6]

---

[3] See E.T. Bell, *The Development of Mathematics* (New York: Dover, 1972), 145–154.

[4] See Appendix A and Appendix B on pp. 16–17 above.

[5] I comment on my experience using heuristics with undergraduates in James Duffy, "Lonergan Gatherings 7: Words, Diagrams, Heuristics," accessed June 1, 2022, http://www.philipmcshane.org/lonergan-gatherings.

[6] The diagram "Structured Wonder" in Philip McShane, *Wealth of Self and Wealth of Nations: Self-Axis of the Great Ascent,* 2nd ed., ed. James Duffy (Vancouver: Axial Publishing, 2021), 14. The first edition of this book is available online at: http://www.philipmcshane.org/published-books.

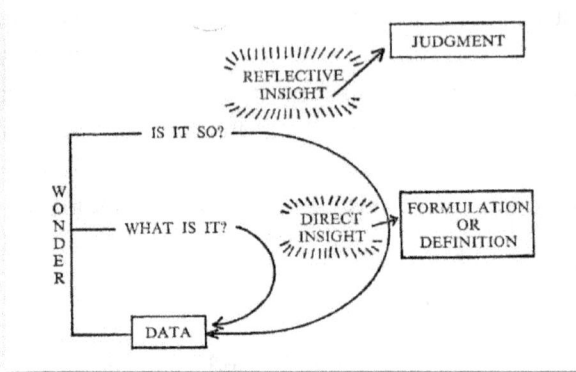

Notice in the diagram and, more importantly, in your daily questings, that what-questions are distinct from is-questions, something McShane alludes to in "The Noosphere" in his commentary on J. Mohorčich, "Scale, Noosphere Two, and the Anthropocene." Notice too that I am following my second piece of advice regarding diagrams.

There is a difference between feeling blue (data), on the one hand, and asking why and what to do about it, on the other. Feeling blue and whatting about it are both conscious activities, thus more like indigestion than digestion. But experiencing the feeling is only a condition for the occurrence of whatting. I might feel blue all day long but never pause to wonder why.[7]

There exists the possibility, beginning in early education and continuing all the way to leisurely sabbaticals and beyond, of noticing whatting, appreciating whatting, cherishing whatting, patiently exploring the role of what-questions in human living, and learning to live them. I am reminded of Socrates. I am also reminded of Rilke writing to a young man:

> You are so young, so before all beginning, and I want to beg you, as much as I can, dear sir, to be patient toward all that is unsolved in your heart and to try to love the *questions themselves* like locked rooms and like books that are written in a very foreign tongue. Do not now seek the answers, which cannot be given you because you would not be able to

---

[7] With patient reflection one might find five distinct answers—classified by some as "material," "formal," "efficient," "exemplary," and "final causes"—to five variations of the question "Why is the sky blue?"

live them. And the point is to live everything. Live the questions now. Perhaps you will gradually, without noticing it, live along some distant day into the answer.[8]

Paying attention and patiently self-appropriating whatting is regularly and sadly not a priority in education in the negative Anthropocene, which finds us with heads and hearts partially if not totally eclipsed. Instead of exercising to avoid academic sluggishness, we have been bamboozled by the "arrogance of omnicompetent common sense"[9] into memorizing names. My role as teacher is to help learners appreciate the difference between seeing—for example, round things at home and at school—and understanding—for example what a *circle* is. But I cannot help them if I myself am not appropriating my whatting adventure, my quest in the mess of the negative Anthropocene.

---

[8] Rainer Maria Rilke, *Letters to a Young Poet*, trans. M. D. Herter Norton (New York: W. W. Norton, 1934), 34–35.

[9] Bernard Lonergan, "Questionnaire on Philosophy: Response," in *Philosophical and Theological Papers 1965-1980*, ed. Robert C. Croken and Robert M. Doran, Collected Works of Bernard Lonergan 17 (Toronto: University of Toronto Press, 2004), 370.

III. Beginning Economics

Economics in the Anthropocene: Blue, Green, and Other Colors

Terrance Quinn

This essay partly concerns Blue Economics, Green Economics and Ecological Economics. I am also thinking of the group Economics for the Anthropocene, Regrounding the Human-Earth Relationship.[1] Blue and Green economics look more to political action, while Ecological economics looks more to academic questions. Generally, though, we find a consensus that in order for economics to be sustainable, it at least needs to not destroy our world's ecosystems. The project called Economics for the Anthropocene (E4A) contributes by being "a graduate training and research partnership designed to improve how the social sciences and humanities connect to ecological and economic realities and challenges of the Anthropocene. Overarching goals are to articulate, teach and apply a new understanding of human-Earth relationships grounded in and informed by the insights of contemporary science."[2]

These groups share a desire and willingness to take action and resolve crises, especially ecological crises linked to economics. So far, the context of discussion about the economics part of the equation is the current standard model that is taught in schools and universities around the world.

An aim of ecological economics is economic development that does not undermine ecosystems. But, what is meant by *economic development?* On present one-flow models, economic development is defined and measured by GDP-type metrics along with indices of world-casino stock markets. This blog is no place for analysis. But, a question fits: Is it not at least a little bit puzzling that for more than a century, in all major economies of the world, GDP based metrics as well as stock market indices have been mainly rising and yet there has been no commensurate climb in economic well-being for peoples of the world? Does this not suggest that there might be something wrong with the current standard model(s)?

What is the source of the present standard model? Are not we—"we" in the sense of whoever is in economics—the source? Does not our thought in economics emerge from our inquiry in and about economics? As it happens, today's what-ing in economics mainly is about speculative models that do not bear

---

[1] "Economics for the Anthropocene: Regrounding the Human-Earth Relationship," Economics for the Anthropocene (E4A), accessed June 1, 2022, https://e4a-net.org.

[2] "What Is E4A?," Economics for the Anthropocene (E4A), accessed June 1, 2022, https://e4a-net.org/what-is-e4a.

out in actual "human-Earth relationships grounded in and informed by the insights of contemporary science."[3]

Fresh growth is needed. As per the advice of my colleagues, that partly will be through a "regrounding" possible by what-ing about our what-ing in economics. If that what-ing about what-ing includes what-ing about actual economic transactions, we will be on our way toward a new science of blue, green and ecological economics. In that new science, the words 'blue,' 'green' and 'ecological' will be unnecessary. For, that new science will normatively be an eco-friendly collaboration, wherein potentialities of nature make possible spectra of flows and standards of living.[4]

In his poem "Of Mere Being," the American poet Wallace Stevens wrote "The Palm at the End of the Mind. ...The wind moves slowly in the branches. The bird's fire-fangled feathers dangle down." In the new two-flow economics, we may find the palm at the end of the mine. And, as we grow in what-ing about our what-ing about our what-ing in economics, we will be finding our way into economics for the positive Anthropocene. Thoughts on focused beginnings will be left for another essay.

---

[3] "What Is E4A?"

[4] For references and a general audience introduction to two-flow circulation of monies in actual economies, see Quinn, "Anatomy of Economic Activity."

Strange Business in the Anthropocene

Terrance Quinn

strange: not native to or naturally belonging in a place: of external origin, kind, or character [1]

In their multi-media website, *The Anthropocene Project*,[2] Burtynsky, Baichwal and de Pencier provide compelling images of what we are doing to our world. Ongoing reports from environmental sciences leave no room for reasonable doubt: the way that we are living is undermining our world's ecosystems and jeopardizing our survival.

What is also known is that one of the reasons for this crisis is an "absolute contradiction"[3] between, on the one hand, endless capital expansion and maximization of profits and, on the other hand, the needs of humanity and ecological sustainability. Elsewhere on this site, my colleagues raise foundational questions about the Anthropocene Epoch. In this note, I look to a less fundamental issue, but one that is altogether urgent. A main purpose here is to invite progress toward better understanding the global financial system. While not the only factor, it is contributing to the crisis in a big way and is a system that, at present, happens to be controlled by less than 1% of the world's population.[4]

---

[1] In *Merriam-Webster*, n.d., https://www.merriam-webster.com/dictionary.

[2] Nicholas de Pencier, Edward Burtynsky, and Jennifer Baichwal, "The Anthropocene Project," The Anthropocene Project, accessed June 1, 2022, https://theanthropocene.org.

[3] Toni Ruuska, "Capitalism and the Absolute Contradiction in the Anthropocene," in *Sustainability*, 51–67.

[4] For instance, of that elite group, "42 individuals have as much wealth as the bottom 50%" of the world's total population (Joseph E. Siglitz, "The American Economy is Rigged," in *Scientific American, Policy and Ethics*, September 1, 2018, https://www.scientificamerican.com/article/the-american-economy-is-rigged). We are on track for less than 1% of the population to own more than 60% of the world's wealth, by 2030 ("Richest 1% on target to own two-thirds of all wealth by 2030," https://www.theguardian.com/business/2018/apr/07/global-inequality-tipping-point-2030). See also, *ChicagoBoothReview*, http://review.chicagobooth.edu/economics/2017/article/never-mind-1-percent-lets-talk-about-001-percent.

Reports from Stock Markets

To observe the financial system in action, we might well start by looking to the stock markets. Daily and weekly reports can be dizzying: "Dow plunges more than 800 points in worst days since February. Amazon shares lead the route."[5] "Rising rate fears and pivot out of technology stocks have made it a rough last few days. The Dow has dropped four of the last five sessions."[6] These are but two of a vast array of financial news reports from October 2018. Earlier that same month, the news was quite different. "Dow climbs 160 points as Troubles melt away."[7] And, of course, similar ups and downs are reported each day and week.

Mechanisms of Stock Markets

In the New York Stock Market alone, billions of trades are made each day. Some buy-and-sell strategies are based on systems of partial differential equations. There is, for instance, the famous Black and Scholes model for pricing options. Computerized models are used to simulate buy-and-sell patterns. Computations and simulations are for analyzing past trends and for working out possible future trends in frequencies and prices. But, of course, even when "risk" (as defined in the various models) is thought to be low, trends in actual market values can and usually do differ from simulated trends.

The name of the game is profit; and profits are made by successfully anticipating either climbs or drops in market values.[8]

Large simulations are run with computers. But computers are also used in another way, as surrogate traders. I am referring to *algorithmic trading* and "black box trading." In investment firms, computers are set up to receive updates on market values. Trades are triggered when prescribed thresholds are met. Is this unusual? It is estimated that more than 40% of the world's trading volume is now algorithmic. In the USA, that fraction is closer to 70%. In other words, large percentages of trades in the world's stock markets now occur due to computer programs "competing" against competitors' computer programs, in a context of

---

[5] www.cnbc.com, October 10, 2018.

[6] www.cnbc.com, October 10, 2018.

[7] Crystal Kim, "Dow Climbs 160 Points as Troubles Melt Away," in *Barron's*, https://www.barrons.com/articles/dow-climbs-160-points-as-troubles-melt-away-1538585975, October 3, 2018.

[8] Investors appeal to numerous factors including, e.g., estimates of short-term and long-term profitability of a company listed on the stock market. An elementary instance of *market value* is share price times the number of shares available. *Book value* is the total value of assets minus liabilities, also known as *shareholders' equity*. Book value and market value usually differ, a difference that is systematically exploited by traders.

market trade patterns that, likewise, is to a large degree determined by large-volume computer generated trades. A common type of algorithmic trading is called *high frequency trading*, wherein batches of trades occur in seconds and milliseconds.

Could the disconnect between stock market strategies and needs of economies, humanity and world ecologies be more obvious?

Today's Large Corporations: An Example

In 2013, Sears Canada was in financial decline. Over the next few years, approximately 509 million dollars were paid to shareholders, funneled through a US Hedge Fund. After filing for bankruptcy in 2017, executives received more than nine million dollars in bonuses.[9] The company then laid off 17,000 employees with no severance pay. Legalities aside,[10] the basic approach is consistent with contemporary principles of "good business." Similar examples are commonplace in world business news.

Global Stock Markets: A Worldwide Vascularization

That stock markets operate like casinos might not in itself be cause for concern. But the stock markets of the world are not merely casinos doing their own thing. Having commoditized monies, the Global Casino is tied into and fed by the vast financial resources of the world's economies. Whether *boom* or *bear*, effects do not stop at the Wall. Indeed, the ups and downs of Wall Street and other stock markets can send shock waves into the world's economies and in some cases have led to economic disaster on a global scale. There was, for instance, the "2008 meltdown," effects of which are still being felt ten years later.[11] To add injury to injury, in periods of financial crisis caused by these activities, standard mechanisms

---

[9] https://globalnews.ca/news/3598469/sears-canada-lay-offs-management-bonuses

[10] Legal battles are ongoing. See, e.g., https://www.cbc.ca/news/business/sears-canada-eddie-lampert-dividends-shareholders-1.4896425. Employees have filed a class action suit against Sears.

[11] John W. Schoen, "Financial crisis of 2008 is still taking a bite out of your paycheck 10 years later," https://www.cnbc.com/2018/09/11/financial-crisis-of-2008-still-taking-bite-out-of-your-paycheck-report.html. See also, Lydia DePillis, "10 years after the recession began, have Americans recovered?" https://money.cnn.com/2017/12/01/news/economy/recession-anniversary/index.html.

allow investors to safeguard their winnings by moving them to protected domestic and offshore accounts.[12]

Shareholders Share Not

Decisions of a CEO and corporate board members are for serving the interests of shareholders; and shareholders generally want to increase profits. What happens to those profits? Even for the "super-rich," there are only so many homes, boats and expensive cars that one can own and use in a year. In large measure, profits are channeled back into new capital expansion opportunities for more profit; into the Global Casino, also for more profit; or else are moved to protected domestic and offshore accounts where they are both out of harm's way as well as out of being able to work in the world's economies. The Shark Tank mentality openly ignores the needs and concerns of more than 99% of the world's population. As we are witnessing on a global scale, nor does an enshrined profit motive require that CEOs, board members and shareholders bother to anticipate ecological ramifications of their decisions.

A World Monkey Puzzle

Like the tree of the same name,[13] world economies and ecologies are endangered, in this case, partly by a parasitic Global Casino that, together with large corporations, operate with open disregard for employees, humanity and the world. However, the tree has other functions. For instance, sometimes changes of ownership make good sense; mechanisms of international banking are useful in global supply chains; currency exchange firms play a role in providing financial aid in philanthropic efforts; sometimes shareholders help move a corporation to solve important problems illustrated, for example, by pressure recently exerted by shareholders on Royal Dutch Shell, to take action on climate change; and so on.

I ended the last paragraph with "and so on." I have provided only a few superficial observations. We are struggling with a global crisis that is partly the result of strange and inimical agendas operative in today's global financial systems. It would be good to understand and be able to distinguish pathological from normative structures. Processes involved, however, are massively complex. What would be helpful is a collaborative effort to understand, in detail, goals, methods and mechanisms of the Global Casino and standard business practice.

---

[12] This is normal and legal in many countries. It is estimated that the present value of offshore accounts is more than 27 trillion dollars and is increasing annually by 5% or more.

[13] Monkey puzzle: *araucaria araucana.*

Toni Ruuska's Capital Ideas

Terrance Quinn

It will help to start by saying something about the title, "Toni Ruuska's Capital Ideas."[1]

I am using the word 'capital' in more than one way. I mean it as an adjective. Ruuska's article concludes by expressing the need "to find fundamentally different ways of being that consider both inter- and intra-species questions to peaceful coexistence on planet Earth."[2] That is a capital idea.

My essay also is to help us "make capital out of"[3] Ruuska's capital idea. I take his plea seriously, that our work needs to do more than contribute to "PhDs, articles and books about these issues."[4] So, I am trying to do some small part toward helping us make progress, together, from Ruuska's paper.

Of course, in Ruuska's paper, there are other uses of the word 'capital,' such as *capitalism* and *capital* in economics.

To help us make *progress*? The name 'progress' doesn't really matter here. Whatever we call it, I am finding the heuristics for cyclic collaboration mapped out by our senior colleague[5] increasingly plausible, feasible and compelling. Perhaps you will begin to as well?

I have been reading and re-reading Ruuska's paper, picking up on various aspects that I think warrant further inquiry. Using the neologism from "Crecycling *Sustainability*,"[6] I am attempting a *crecycling* of his article.

> The watching, normatively, is caring. But in this first Næssian "division of labour" its care leans forward within whatever Model is available—glorious in physics, shabby in economics—to detect, inspect, the vibrant data for nudges towards progress. We can comfortably call this detecting *Research*.[7]

---

[1] Ruuska, *Sustainability*, 51–67.

[2] Ruuska, *Sustainability*, 64.

[3] "to make capital out of," idiom, to use (something) to one's (our) advantage (https://www.merriam-webster.com/dictionary/make%20capital%20out%20of).

[4] Ruuska, *Sustainability*, 64.

[5] Philip McShane, "Crecycling *Sustainability*."

[6] McShane, "Crecycling *Sustainability*," 89.

[7] McShane, "Crecycling *Sustainability*," 91.

I recall this sentence from "Crecyling *Sustainability*" because, in my reading of Ruuska's work, it seems to me that there are indeed nudges towards progress. But it won't be much help if I simply repeat Ruuska's words. Instead, I make an effort to draw attention to aspects of his work that to my mind invite *Interpretation*.[8]

I have been finding this effort to *creatively recycle* both interesting and challenging. In hindsight, I am reminded of teams who work at CERN, for example. Researchers pore over photo-plates and sift through numerical data, in an attempt to find variations that might be significant for progress. I have been poring over Ruuska's paper. At this point, there are not yet teams of collaborators devoted to this kind of work. I have done what I can and have been finding aspects of the article that, to my view, call for further attention. Call for further attention? In the sense that they might be significant for progress and, toward that end, I share them with you as best I can.

But I also have been finding aspects of the paper that, to my view, are problematic that, in analogy with physics, reveal a kind of "interference." As I will get to below, I suspect that drawing attention to that interference also can help in our effort to make progress.

I also note that by trying to be as up front as possible about what I am doing, data might be more easily retrieved later by someone else, perhaps the reader of this essay, someone working on different problems or someone with a more advanced view. My hope is that my sifting and sorting might be of help to later Researchers, Interpreters, Storycheckers, and so on.[9] My main "hand off," though, is to Interpreters. Sill, my effort here is preliminary. It may not yet be very helpful for Interpreters. In that case, maybe it will be worth crecycling.

I wondered whether I should start by drawing attention to what I think are problematic aspects of Ruuska's paper or start with those aspects that more readily invite Interpretation. In the end, I decided to start with the former. That way adds context to my discussion. Also, as I explain below, I will not attempt to address problematic issues in this essay. Why not? The main aspects that I find problematic are not unique to Ruuska's paper. They are, instead, part of a tradition of philosophic debate that, at this time in history, has been normalized.[10] And so

---

[8] See note 7.

[9] See note 7.

[10] Bernard Le Bovier de Fontenelle (1657–1757) helped establish the tradition of philosophic debate. He wrote for the "worldly salons, …, (whom) he regarded … as his essential audience." Steven F. Rendall, "Fontenelle and his Public," *Modern Language Notes* 86 (4) (1971), 496–508.

it seems to me that any attempt to seriously discuss those issues in a helpful way would be too much for a review essay, let alone any single scholarly article. What I perceive to be problematic features of Ruuska's paper will, I think, be better handled by those who are focusing on questions such as "How are we doing and How are we doing in asking How are we doing?"[11] In other words, I do draw attention to aspects of Ruuska's paper that I think are problematic. But I don't attempt to contribute to the work of resolving those issues. Instead, I pass them forward to whoever happens to be focusing on more fundamental questions, to those involved in a fourth task indicated by 'D' in the figure given at the end of "Steps Nine and Ten."[12]

Which aspects of Ruuska's chapter do I find problematic?

> Hence, the task of defining capitalism is difficult, and consequently it is no wonder there is lack of consensus regarding what capitalism is ultimately made of (Graeber, 2014). Yet, in order to understand or to critically analyse capitalism, there arguably needs to be some kind of definition of it.[13]

Ruuska goes on to add his own definition to the mix: "Thus, to summarize (see Table 4.1) and elaborate still further, capitalism is a historical socio-economic structure in constant and endless pursuit of accumulation of capital."[14] Is the new definition better than previous definitions? How does it relate to prior definitions? But really, the problem is more fundamental, as can be witnessed in the bent of the argument. In a context that is partly determined by the new definition, the argument speaks of "rentiers," "capitalists," and so on. But, who, in particular, are these rentiers and capitalists? What does an argument about "some kind of definition"[15] tell us about actual goings on in particular homes, villages, cities, about the activities of the (much less than) 1% that has representatives in most major economies, stock exchanges, offshore banking and other goings on in actual global economics?

There is also the fact that the analysis[16] partly depends on numerous quotations of (apparently) supporting voices. Are they actually supporting voices? Are there, perhaps, other voices that might also be relevant? Are the meanings of

---

[11] McShane, "Crecycling *Sustainability*," 93.
[12] McShane, "Crecycling *Insight*," 104–107.
[13] Ruuska, *Sustainability*, 52.
[14] Ruuska, *Sustainability*, 54.
[15] Ruuska, *Sustainability*, 52.
[16] See note 13.

authors quoted obvious, simply by reading their words? Might it not be convenient, perhaps even crucial, to be able to appeal to interpretations of important authors cited? This may seem to be asking too much of one paper. And, it is. What I am getting at is that what is being revealed are problems intrinsic to a philosophic tradition that does not effectively distinguish different tasks.

With that said, Ruuska's context of concern clearly goes further than philosophical debate and the words of other authors. He speaks of the Industrial Revolution, surging population groups, technologies, "our precious planet,"[17] "global-scale problems concerning natural environments,"[18] the market economy, some of today's "key institutions,"[19] and much more. And, as already mentioned, the paper ends with his capital idea.

It seems to me that one way to begin to get a hold of positive content of Ruuska's article is to temporarily replace a word in his discussion with a place holder. Throughout his paper, he describes extensive damage being done by institutionalized greed that is part of today's establishment economics. Replacing the word "capitalism" with the name "today's establishment economics" leaves most of his descriptions intact, preserves his evident concern for what is in fact going on, is compatible with his desire to draw attention to a fundamental incompatibility of today's standard economic practice with long-term sustainability (and peaceful coexistence), but also helps filter out, or at least hold back, interference coming from philosophical argumentation about a pre-defined "capitalism."

Each paragraph of Ruuska's paper mentions something about economic principles, economic practice or both. With a focus now adjusted more to what in fact is going on, two questions are:

(1) How does an economy work?; and
(2) What does Ruuska mean by 'economy'?

These are the two questions that I hand forward for Interpretation. They both emerge from Ruuska's paper. They are both key questions, for progress with either or both would help us move forward in our search for solutions to economic problems in the Anthropocene. It is true that both questions will be enormously difficult. And, yet, clues for both are available in Ruuska's article.

---

[17] Ruuska, *Sustainability*, 51.
[18] Ruuska, *Sustainability*, 51.
[19] Ruuska, *Sustainability*, 55.

Regarding (2): To make progress in being able to explain Ruuska's meaning for 'economy,' we will need a science of interpretation that does not yet exist. But, already evident, descriptively, is that Ruuska combines everyday description of economic activities with terms such as *production, commodity* and *investment* and so on, terms that are part of today's establishment economics.

Regarding (1): How does an economy work? What is the normative structure of an economy?[20] It seems to me that here too preliminary evidence is pointed to in Ruuska's paper. As he mentions, there is "production of commodities (and there) is production of commodities in order to produce."[21] Within the fabric of his discussion, he ties these to "capitalism." But what I find interesting is that the distinction that he makes would seem to be normative. Can there be an economy that does not produce? In any economy beyond the most elementary fruit and nut gathering (e.g., no baskets yet), are not tools and such produced in order to use in production? In other words, Ruuska's observation would seem to be of something true of any economy. However, what he observes does not fit contemporary establishment theories that rise on descriptive one-flow models about "homes and businesses." He again touches on the distinction when he writes of "consumable goods" and "some fraction of the production is more capital" (goods).[22] He also observes that while these two types of economic activity are distinct, they are also linked in various ways. For instance, he speaks of wages (not spelled out but, implicitly, in both), of increasing (or decreasing) output, and more, all of which depend on financial flows within both types of production and between both types of production.

The help Interpreters, I give a preliminary diagram for these distinctions, namely, that in any economy there seem to be two kinds of production:

---

[20] In addition to normative structure, there is also the problem of cultural ethos. The normative structure is present whether an economy is under the control of a sociopath who imprisons their own people, or an economy is providing an abundance of goods and services to a diverse and thriving population. The problem of ethos is a topic in, Jessica Lawrence, "Managing the environment: neoliberal governmentality in the Anthropocene," chapter 5, in *Sustainability*, 68–84. I discuss Lawrence's article in "Managings of History: Governings in the Positive Anthropocene," pp. 59–64.

[21] Ruuska, *Sustainability*, 52.

[22] Ruuska, *Sustainability*, 57.

Production for production

?   ?   Production

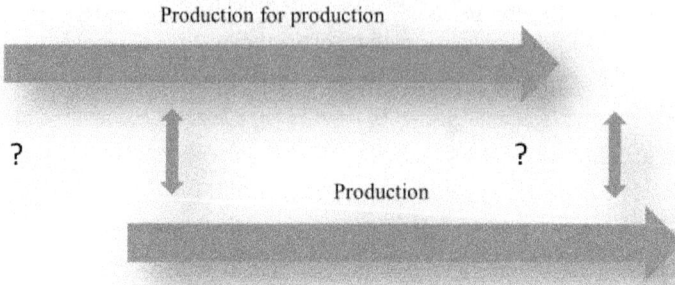

Question (1) for Interpretation would seem to call for developments in understanding how these two types of production are linked in actual economies of the world. As Ruuska's paper brings out, we would also eventually need to account for financial sectors.

IV. Shifting Probabilities

It's Getting Better and Better, Worse and Worse

Philip McShane

Todd LeVasseur's chapter 6, "'It's Getting Better and Better, Worse and Worse, Faster and Faster': The Human Animal in the Anthropocene" is a curious mixture. There is a first interest in the emergence of humanity, as it is described by K. Sale, with its focus on caloric intake leading to reflection on humanity's narrow localization in icy times and a concomitant surge in the invention of weaponry 10,000 years ago. Then the interest moves to what I might call a weaponization of counter-weaponry, where now the weapons to be countered include the large-scale assumptions and structures of modern so-called civilization: that weaponization is sufficiently conveyed by the title *Deep Green Resistance* especially in its association with *Deep Ecological War*. On these I commented at the end of "The Noosphere." And in that conclusion, I pointed towards a quite new theoretical background needed to get beyond such resistance and warring. LeVasseur, of course, is not talking of that new perspective when he talks of thinking out a conceptual background. He is referring to Sale when he writes, "this gives a theoretical background to the larger reality that we must face if we wish to understand what preconditions are needed to have a more peaceful coexistence in the Anthropocene."[1]

The drive towards this peaceful coexistence must "address possibly deep-seated misothery and the manifest cultural institutions and technologies that have been scaffolded upon such misothery, with this scaffolding stretching back thousands of years."[2] The scaffolding and the misothery have a genetic history of at least 22,000 years. Such, sweepingly, is the diagnosis over which "we must pause as scholars (and as citizens of a beleaguered Earth)."[3]

LeVasseur goes on to the issue of prescriptions, but his focus is more on understanding than action. Sale "can help us to understand"[4] but the poise is on coming to grips with the past and that effort is only thinly prescriptive: to this I return suggestively in the conclusion So, we arrive at the second section of the essay, titled, "Discussing the Anthropocene, or cultivating 'ecological thought,'" where LeVasseur begins by talking of hope, but is not effective in generating such

---

[1] *Sustainability*, 90.
[2] *Sustainability*, 90.
[3] *Sustainability*, 89.
[4] *Sustainability*, 89.

hope: "even discussing the term 'Anthropocene' is no easy task"[5] and his pointers here to such ventures as analyses of meta-narratives are of little hopeful help.

In section 2 the focus is on "hyperobjects," a word coined by Morton, a distracting word that I would suggest you replace by the German *Praxisweltanschauung*, a pragmatic boost of the German word for worldview. If the view is objective than yes, there is a reality, an object, like human dominance over soil. But what is important is to get the view that a view is the core of the problem. Yes, humans are "embedded within hyperobjects": viewpoints or their objective geohistorical complexes. The present sick embeddedness "confronts us with the chance to question what it is to be human,"[6] to push "a process of becoming fully aware of how human beings are connected with other beings—animal, vegetable, mineral. Ultimately, this includes thinking about democracy."[7] All this is the goal of ecological thought.

LeVasseur is right on here, but the push, the confronting, fades in his next pages. To that, I return in the conclusion. And to get there I need only note that in section 3, Deep Green Resistance, he returns to puzzle further over aggressiveness as illustrated by the Deep Green Resistance movement, and in the final section 4, "Citizenship in the Anthropocene," he poses a central question of human hope, a "blessed unrest"—to quote Paul Hawken: "the call that we redefine our citizenship."[8] I am reminded of an *Essay on Fundamental Sociology*, written by Bernard Lonergan during the Italian and German nationalisms of the 1930s, where he remarks on "the stupid appeal to a common language and a united geographical position as something of real significance."[9]

So, yes, we need a move to redefine our citizenship, the move to invent a hyper-object and hyper-objective: a *Praxisweltanschauung* that is first a possession of the human mind and then battles its way to global tranquility. Will the battle involve real clashes of protest groups with the military victim caught in uniformed allegiances to present pundits? That is a serious question of this next century, and the answer, whether hyperobject or hyperobjective, cannot but be a mess, a mess to be lifted into a *Praxisweltanschauung*.

---

[5] *Sustainability*, 91.

[6] *Sustainability*, 93.

[7] *Sustainability*, 92.

[8] *Sustainability*, 98.

[9] Michael Shute, *Lonergan's Early Economic Research* (Toronto: University of Toronto Press, 2010), 30.

Morton, quoted by LeVasseur, poses the question of that lift: "what would a truly democratic encounter between truly equal human beings look like, what would it be—can we even imagine it?"[10] We are back, or forward, in a new context, to the issue of the question posed, the question marked, the last and the first 'words' of the quotation from Morton that is, sadly, the neglected beginning: it? : what. But I do not wish to repeat my earlier reflections on chapter 7 of the book. Rather I would have us home in on the question LeVasseur raises, regarding finding a framework that would lift probabilities of peaceful coexistence. Let us return, then, to his considerations of the value of Sale's work. What is the basic pattern of that work? It moves, as indeed the science of geology does from data of the past, through efforts to interpret that data, to the matter of checking or verifying: to this there is added such reflections as LeVasseur, on the value and direction of the work.

We have here, then, something quite general—think of physics characterized as the "verification of theory in instances," a three-layered strategy that, in our times, regularly requires massive division of labor: think of the hunt for the Higgs particle. Evaluating progress is evidently an added process. "How are we doing?" is a simple aspect of that evaluation, and the simple aspect brings out a four-layered process as bent on ordering the past. Old-style geology required research within a theoretic that led to verification: and the dynamics of that triple process was wisely evaluated as it emerged. But the process was spontaneous, as was a parallel disorganized forward-looking triple process of policy, planning, and the creation of further structured data. So one finds a division of labor that is fundamental and eightfold.

This discovery is not new. It is to be found in the writings of the father of "deep ecology," Arne Næss, whose search led him to the structuring of suggestions of Gandhi. He wrote thirty years ago of "Rethinking Man and Nature: Towards an Ecological Worldview."[11] It reflected his original deep ecology paper of 1972–73, referred to in my concluding quotation. It is sad that the writers of this volume on *Sustainability* did not advert to his tentative pointings towards an eight-fold division of labor. It is painfully evident that the sincere but scattered efforts of these dedicated authors and of the whole effort of the Anthropocene movement, with or without a capital "A," would benefit from Næss's *Praxisweltanschauung*. His view pointed to structures of data and of collaboration,

---

[10] *Sustainability*, 93.

[11] Arne Næss, "Rethinking Man and Nature: Towards an Ecological Worldview," *The Ecologist* 18 (1988), 118–185.

and he pushed on to identify an overarching hyperobject: "Applied to humans, the complexity-not-complication principle favours division of labour, *not fragmentation of labour.*"[12]

---

[12] Næss, "The Shallow and the Deep, Long-range Ecology Movement. A Summary," 97.

Managings of History: Governings in the Positive Anthropocene

Terrance Quinn

> "So that, in the end, there was no end."
> –Patrick White, *The Tree of Man*.

Ten heads are better than one

Topics in Lawrence's paper[1] include, among other things, "neoliberal governmentality," "law" and the "Anthropocene." It seems to me that throughout the paper a dominant question-mode is "How are *we* doing?" (where *we* refers to "human individuals and societies,"[2] globally). Based on detailed observations about EU Environmental Policies from 1973 onward, Lawrence ends with two paragraphs that, among other things, point to the need for "reimagination of our logic of government"[3] as well as the need for us to "break free of and avoid being coopted by a political force that has gained increasing control over our human imagination: neoliberal."[4]

How will I follow up on Lawrence's suggestions, in a way that might contribute to our common goal, that might help *us* ("human individuals and societies") make progress? I am, then, thinking like a collaborator but, in this instance, not just any kind of collaborator. In Lawrence's paper, we have a nudging to meet an increasingly evident need for a new "human social order."[5] However, her conclusions are open ended. As she writes, there is a need for "truly rethinking the human relationship with the planet"[6] and for "reimagination."[7] Also, there seem to be fundamental differences between her views and mine. That isn't a bad thing. I have learned from Lawrence's paper and found that my view has been enriched in the effort. Nevertheless, if there is "hope"[8] for being able to join forces in a way that might help "individuals and societies," there are difficulties that will

---

[1] Jennifer Lawrence, "Managing the Environment: Neoliberal Governmentality in the Anthropocene," in *Sustainability*, 68–84.

[2] Lawrence, *Sustainability*, 68.

[3] Lawrence, *Sustainability*, 82.

[4] Lawrence, *Sustainability*, 82.

[5] Lawrence, *Sustainability*, 82.

[6] Lawrence, *Sustainability*, 70.

[7] Lawrence, *Sustainability*, 82.

[8] Lawrence, *Sustainability*, 82.

need to be addressed. For instance, both Lawrence and I use the name *Anthropocene*. But what do each of us mean by the name? If we have very different meanings then, without taking some care, discussion could easily generate misunderstanding rather than help us make progress together.

In fact, it seems to me that Lawrence and I do have different meanings for the name *Anthropocene*. Our academic backgrounds are also different and so too are our horizons. What are *we* to do, where *we* would include Lawrence and me, but also anyone interested in contributing to the present task? Might we not be helped by each of us attempting to articulate aspects of our view that seem especially pertinent to the present context, as well as how we reached our views? That would be a personal achievement. But as often happens quite naturally in conversation, more is possible. In addition to attempting to make sense of what one means, and what our collaborators mean, why not also check with our collaborators to get feedback on our view and on our view of our collaborators' views?[9]

My View of Lawrence's View

Lawrence's paper includes discussion of various conceptual constructs. For instance, in addition to "neoliberalism" there is also "Anthropocene": "Part of the Anthropocene's conceptual weight stems from the challenge it poses to the modernist idea that there is a firm conceptual boundary between nature / environment/non-human and culture/constructed/human."[10] But it seems clear to me that Lawrence's context of concern reaches well beyond such boundaries and includes actual goings on in law, governments, "Earth systems," management, as well as an awareness of an increasing "extension of the logic of the economy to the management of all areas of life."[11]

Whatever one's preferred definition of "neoliberalism,"[12] Lawrence goes on not to logical debate but to describe and discuss aspects of contemporary circumstances. She points to a deepening grip of "free market political

---

[9] In incomplete and non-technical fashion, I am touching on a few aspects of the task called Dialectics. See note 47.

[10] Lawrence, *Sustainability*, 68.

[11] Lawrence, *Sustainability*, 68. Much as I suggested in the essay on Ruuska's paper, to help filter out conceptualist influence and that way to better highlight Lawrence's overriding interest in and concern for actual goings on, in her paper we can, for instance, replace "neoliberalism" with "the contemporary ethos."

[12] "Definitions of the term 'neoliberal' differ (Peck, 2010)" Lawrence, 69.

philosophy"[13] that includes an increasing influence of "strong property rights, free market, and free trade."[14] She draws attention to a contemporary ethos that increasingly "makes the market (rather than 'rights' or 'freedom,' for example) the regulating principle that guides the actions of the state,"[15] that "seeks to apply economic analysis[16] and market techniques to all areas of life,"[17] whose laws are "made not by classical jurists concerned with right and the limits of power, but by experts, technocrats and economists concerned primarily with the efficient and effective management of a 'marketized' social order."[18]

By appealing to examples,[19] Lawrence goes on to highlight the fact that the "Anthropocene as a concept does not have the revolutionary impacts that many scholars have hoped for."[20] A main point of her paper, she draws attention to the fact that current talk about the Anthropocene has emerged from within, and remains part of, the current ethos. She sheds further light on the fact that, in recent decades, there has been a transitioning toward a "marketized social order." And so: "If there is any hope for truly rethinking the human relationship with the planet, therefore, it will require more than simple recognition of our role in shaping it: it will require a concerted resistance to and reimagination of our logic of government."[21]

I find it helpful to group some of Lawrence's suggestions as follows: (a) There are broadly stated hopes: Lawrence writes of the need for something "revolutionary;"[22] that we obtain a "more harmonious relationship with Earth systems"[23]; that we live in a way that is not "reducing our gaze"[24] and is, instead, one that promotes "expanded consciousness."[25] (b) She also gets more detailed:

---

[13] Lawrence, *Sustainability*, 70.

[14] Lawrence, *Sustainability*, 70.

[15] Lawrence, *Sustainability*, 71.

[16] The reference here is to "economic analysis" in the contemporary flawed standard model economics.

[17] Lawrence, *Sustainability*, 71.

[18] Lawrence, *Sustainability*, 72.

[19] "The rise of neoliberal managerialism in EU environmental policy," section 3, 73–77, and "Managing the environment: the neoliberal Anthropocene," section 4, 77–81.

[20] Lawrence, *Sustainability*, 81.

[21] *Sustainability*, 82.

[22] *Sustainability*, 81.

[23] *Sustainability*, 81.

[24] *Sustainability*, 82.

[25] *Sustainability*, 82.

Aspects of our new way: should include "limits of (economic) growth;"[26] should *not* include an "intended control;"[27] should *not* consist of ways "recognized and accounted for by expert managers;"[28] be so that "law and government gradually come to take up the language of the Anthropocene;"[29] should be so that "political systems will have to respond to the new understanding that human beings are part of, and the most important influence on, Earth systems."[30] And: "*How* this political and legal response will happen is emphatically a function of human social organization."[31]

Our Views and Hopes

Lawrence observes that, as it is so far understood, the "Anthropocene" has emerged within and is part of the current ethos. Recent discussions associated with "Anthropocene" have not yet resulted in significant change; but at the same time, we are in need of something "revolutionary."[32]

In my view, Lawrence makes an important observation. Although, as I will explain, it is an observation that needs to be recast in a new context. Her work is precise, descriptive, scholarly, and connects with up-to-date issues, economic policies and standard practice. I am, however, not seeing evidence of, for instance, influence from the modern sciences nor, for example, up-to-date heuristics for "expanded consciousness,"[33] "function,"[34] "human,"[35] "social"[36] and "organization."[37] Also, (while at this time in history it is to be expected) Lawrence's talk of economics, markets, (economic) "growth" and "limits to (economic) growth" are in terms derived from fundamentally flawed contemporary standard models.

---

[26] As in note 16, "growth" is "economic growth" as defined in contemporary economics.

[27] Control refers to control of our lives by market factors, as described in Lawrence's paper.

[28] *Sustainability*, 82.

[29] *Sustainability*, 81.

[30] *Sustainability*, 81.

[31] *Sustainability*, 81–82.

[32] *Sustainability*, 81.

[33] *Sustainability*, 82.

[34] *Sustainability*, 81.

[35] *Sustainability*, 81.

[36] *Sustainability*, 81.

[37] *Sustainability*, 82.

What I am seeing, then, is that, at least on the surface, Lawrence's paper also emerges from and is part of the current ethos. This may seem to be a surprising claim, for her article explicitly challenges the present ethos.

For me to explain this, I need to say something about my heuristics. I have different heuristics than what is intimated in Lawrence's paper, and different also from what is found in much of the current literature on "the Anthropocene." Like my colleagues contributing to this collection of essays, my view contrasts the negative effects (of the Anthropocene) with the positive effects to come. In other words, I distinguish the present negative Anthropocene from a not yet reached positive Anthropocene.

Some evidence that Lawrence's suggestions do not point beyond the ethos of the negative Anthropocene can be found in (b), above. She gives us glimpses of a suggested new way. But it is a new way that will not be *fundamentally* different. It is to be determined by a new set of laws. It is to be when "the idea of the Anthropocene … begin(s) to enter the law."[38]

On the face of it, this does not mesh with my view of where the whole show is going, where the Anthropocene is going. It isn't possible for me to briefly communicate my present heuristics. However, I can at least point to a range of experience that has contributed to my heuristics, a heuristics of growth. It is by then adverting to a heuristics of growth that I can bring Lawrence's aspirations together with what I envision for the positive Anthropocene.

I devoted decades to learning mathematics. I slowly grew into the tradition, a tradition that has been growing and whose front lines continue to expand more or less exponentially. What a remarkable journey of growth it has been for the mathematics community in history, and—with gratitude to my teachers and collaborators—in as much as I have learned and grown, for me too. There are, for instance, developmental sequences of contexts in history, and now also in me, that begin with Archimedes' discovery of the definition of a limit (of a geometric sum)[39] and eventually reach modern operator theory. That is just one range of growth sequences in history and in my experience.

Lawrence envisions a new social order, but one that will be more or less fixed. If, however, we look not to what she envisions but to her searching, and to some of her observations about EU Environmental Policies, we find not-yet-adverted-to dynamics of growth in her and in societies. So, while our stated views

---

[38] Lawrence, *Sustainability*, 82.

[39] This was his solution to finding the area under a parabola, also known as "quadrature of the parabola."

do not fully coincide, there is a fundamental compatibility in what we do and what we are trying to do: we are human operators of ongoing growth in history.

In fact, is not a heuristics of growth (and decline) also needed for societies, laws and governance? There is, of course, all of history. For one cut, we can read section 3 of Lawrence's paper. She identifies a period of almost 50 years during which EU's Environmental Action Plans gradually moved, in steps, toward a paradigm where policies are now being strongly influenced by market factors and models of "sustainable (economic) growth" (again, where "sustainable" and "economic growth" are defined by contemporary establishment economics[40]). Might not climbing out of the present global mess involve a series of some kind of "recovery steps"?

That last suggestion won't be controversial. But is there to be a final step, a definitive set of laws with a final and universal "logic and government"[41]? Are there prescribed limits to our individual growth, and to growth and progress of societies and cultures in history? Of course, my questions here are rhetorical. In my view, our present negative Anthropocene partly is an ethos that, among its many horrors, is terribly and terror-ably effective in "reducing our gaze to what can be recognized and accounted for by expert managers."[42] It is undermining the possibility of sustainable growth of individuals and societies. On the other hand, my present and preliminary heuristics of the positive Anthropocene are not of a final set of laws or societies. As Lawrence suggests, it will include "a platform for imagining a new human social order."[43] But, that platform also will grow.

What is that platform? In the essays in this collection, my colleagues write not about a pipedream but of a daydream of a "cyclically structured platform" that, in us, in history, has been slowly coming into view in all areas of inquiry. What is (pre-) emergent are eight main tasks, diagrammed at the end of "Crecycling *Insight*."[44] I think now of older meanings of the word 'manage' which include, "to handle," "to touch," "to effect by effort." Transitioning from present struggles to sustainable and peaceful co-existence will depend not on there being "expert managers" (in the contemporary sense of the marketplace) but on the emergence of Increasingly Cyclic Expert Care-filled Managings.

---

[40] See also notes 16 and 26.

[41] Lawrence, *Sustainability*, 82.

[42] *Sustainability*, 82.

[43] *Sustainability*, 82.

[44] See p. 107 below.

Down and Out on Planet Earth

Robert Henman

In 1933 Eric Blair, under the name of George Orwell, published his book *Down and Out in Paris and London*.[1] It was a reflection on the poverty of the masses with a focus on restaurant workers. He did a bit of tramping on his own but left that out of the book. Much of the literature on the Anthropocene reflects on the masses that are today struggling in poverty in order to keep the less than 1% controlling 49% of the global funds while those of us who are middle class do most of the buying. The wealthy buy the big-ticket luxury items. The 'Down and Out' in my context refers to a form of reflection on the economic disparity that leaves one with a negative hope regarding the Anthropocene.

Puzzling away over what I call the negative Anthropocene, with the question "What to do now?" dominates much of the literature on both websites and in books. There are many suggestions offered. Unfortunately, we do not witness positive responses and we might ask why?

Joe Busto, quoted by Barret Baumgart in his *China Lake: A Journey into the Contradicted Hart of a Global Climate Catastrophe*, suggests a reason:

> People just have made their minds up. The paradox seems clear: the more you refute, the more you affirm. Perhaps in such a situation, it's better simply to keep quiet. People know little about cloud seeding because they know little about the natural world, America makes music and movies. It's not interested in science. Climatology, meteorology, hydrology, and resource management—they're all topics that are largely boring outside of the scientist geek world.[2]

But is the public the problem? Governments are asking the public at large to recycle while transnational corporations continue to ravish the earth's crust.

Busto's comments would seem to hold more for the Donald Trumps of the world, and there are many. Governments continue to hold G nation conferences and agree to goals to be met in some distant future. Is there a block to action and if so, what might it be? Ulvila and Wilén in their "Engaging with the Plutocene: moving towards degrowth and post-capitalistic futures" suggest a block to action

---

[1] George Orwell, *Down and Out in Paris and London* (London: Victor Gollancz, 1933).

[2] Barret Baumgart, China Lake: A Journey into the Contradicted Hart of a Global Climate Catastrophe, (University of Iowa Press, 2017), 28.

in question form. "Does the Anthropocene narrative and conceptualization foster the political action needed for tackling the environmental destruction taking place in the Anthropocene era?"[3]

Now, I have no doubt that we all have suggestions as to why implementation of action that would help to meet goals is stalled or inoperative. I have a few of my own. But at this stage I am no more certain of my suggestions than perhaps you are of yours. So, what is the block, and perhaps if we could zone in on that might we glimpse new strategies? Most climatologists and ecologists agree that it is economic activity that is leaving the greatest footprint. Why do the transnational corporations and governments pay little heed to G nation agreements? Well, we do want our Nikes and Toyotas at the cheapest price possible and they know it. So, we are all in the game together. Unfortunately, a large portion of the human population needs to be in poverty in order for us to have such things at the lowest price possible. Is there a need for a new form of education on many fronts that would mediate a more positive force towards action?

Paolo Davide Farah in his Foreword to *Sustainability and Peaceful Coexistence for the Anthropocene* makes the following comment regarding the transformation of the existing governance models into sustainable ones

> that recognize the complexity of relations between the human and non-human worlds. Actually, there is little new to discover. Modern societies should simply learn from the past and implement in a practical manner the fundamental principles and concepts of philosophies from the West and the East that aimed at and envisaged a good balance between humans and nature.[4]

Farah raises the point well in stating that: "Modern societies should simply learn" and that seems to be the crux, we don't learn. Why is that? What keeps us from the type of learning required? Or perhaps better stated: Does the lack of **what** keep us from the type of learning required? Might this be "the little new to discover"? Is there a form of education lacking that would mediate a new form of action? Are we: "down and out on planet earth" because something is missing in our education?

The image of the positive Anthropocene that I brought into this discussion in my earlier piece might have its genesis in locating within ourselves this what-dynamic and all that it mediates. Might the "Great Acceleration" be transposed

---

[3] Heikkurinen, *Sustainability*, 120.

[4] Heikkurinen, *Sustainability*, xvi.

into a greater acceleration of "the eternal strivings of the human spirit towards east, towards Home"?[5]

---

[5] Hermann Hesse, *The Journey to the East* (London: Peter Owen, 1970), 12.

V. Ant Essays

Ant Hop

Philip McShane

The most obvious feature of the shift to the name *Anthropocene* is that it brings into the study of evolution the Anthropos. Less obvious is that it brings into the arena of consideration and deliberation the character of the study itself. For, the introduction of the new name and new period has come forth very centrally because of a heightening of concern regarding the way the emergence of humans has impinged on the evolutionary process. The bent of those concerned with the topic and its real referent is towards a type of study that would shift us out of what we may clearly claim to be the negative Anthropocene, where study has led solidly to what we might call evolutionary ills.

No doubt the emergence of study has brought forth libraries and technologies that somehow can stand apart, in our cities and plains, from their effects on our little planet and its vast environment. Discussion of those effects by ecologists of rural and city structures have, in this past century, pushed us towards an increasingly dark statistics of the balance between what may be vaguely called good and bad effects, consequences. One finds, in some students of these past millennia, a view that would consider putting good and bad in a proportion of, say, 10% to 90%, as altogether too generous. One finds authors who find little to praise in the glory of the Industrial Revolution. Yet that glory is a glory of the type Anthropos, the only extant members of the subtribe Hominina: Homo sapiens. Is the invention of this subtribe, with its academy of studies, perhaps, an evolutionary mistake?

Or is it that the evolution is incomplete, and that its completion is the task of, let us say, the positive Anthropocene?

Mistake, misdirection, deviation, long-run sport: whatever one calls it, there is a growing number of the subtribe, *Homo sapiens*, who sense that something needs to be done about re-direction. Might we go so far as to claim that the evolutionary sport of presently-structured study and indeed its presumed sapience is a death-trail? Could we go annoyingly further and suggest that it has been, yes, the sport of kings and also the sport of saps with sapiens pretensions? Evolution is radically incomplete, headless when it needs a joyful go a head, truncated with a headless body politic presiding over a busy armed decay. So, let us view freshly the beginning, pointing up the thing we name *what*.

> The emergence of humanity is the evolutionary achievement of sowing what among the cosmic molecules. The sown what infests the clustered molecular patterns behind and above your eyes, between your ears,

71

lifting areas—named by humans like Brocca and Wernicke—towards patterned noise-making that in English is marked by "so what?"[1]

"So what?" might be considered as a suitable characterization of truncated humanity's view of the enterprise we are undertaking in this present effort of seeding the positive Anthropocene. Truncated humanity has the word *what* shrunken into a dead noise that in the primitive was a new ape's dark wonder, and my Webster's dictionary weaves its abundant dead namings down a long page to end with "what's what [Colloq.] the true state of affairs."

Yes, "what's what" is the seed now to humanity's identification of the true state of affairs, something to be a luminous identification only in the positive Anthropocene. For now, the task of those interested—concerned, frantic even—about our future, is to continue their varieties of struggles towards cutting down on gross destructiveness and seeding positive creatures, great and small. But the side task that is eventually to lift the small to all is the task of beginning to cherish the colloquialism, What's what.

That cherishing is the core "Opener to the Positive Anthropocene."

That cherishing, when it blossoms in a century or so, will be an open-structured global complex of e-duction, education that will poise all us whats towards "the true state of affairs" in a lean-to poise of what-is-to-be the state of affairs. Perhaps some simple imaging could help us in sensing that strange hopeful poise. Think, then, of the famous Archimedes of Syracuse who died, at 75, in 212 B.C. His 'eureka' story is famous. But now I am thinking of his lean-forward poise when he wished to find a way of getting water to go uphill. Water going up-hill seems a daft proposal. As daft as the proposal of getting culture to go uphill out of the negative Anthropocene. Archimedes' what, in the eureka story, was focused on the present state of affairs: a crown and its content. It is no small challenge to figure out the mind-bump behind the eureka in that story, and we should come to that bump-up-focus later. But here I wish us to cherish in various ways the more strenuous bump-up.

A first cherishing is an astonishment at the odd idea of getting the water to go uphill. Of course, it is not odd if one is thinking of buckets carried and non-leaking carts. But to actually get, so to speak, the stream going uphill?! Archimedes had no thermodynamics, but he knew the uphilling would take work, first his own cultural uphill, then the uphilling of water.

---

[1] McShane, *The Allure of the Compelling Genius of History: Teaching Young Humans Humanity and Hope*, 3.

The second cherishing is an existential pause over the product, without any serious effort to figure out the what, the plan, the state of affairs envisaged: a working reality of watering on higher levels of land. Here you are: a simple image of what is known as *Archimedes screw.*

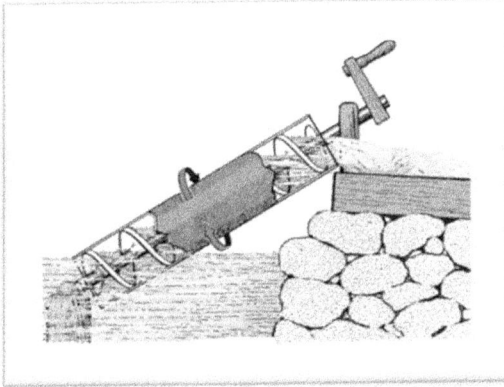

The pause begins with a ramble round the details, with a sense that you really don't know how the wind-up works. But the next step is the mighty one of sensing that there was a more fundamental wind-up in Archimedes that led him, screwed him up to, the plan. Suppose we could diagram that wind-up in Archimedes. Might you come to sense that you didn't really know how the imaged wind-up really works? Our genius paused over one needy state of affairs asking, whatting: "what is to be done?"

The whatting, so so slowly and accidentally, blossoms—eurekas in a massively complex sequence—into, not a bright idea, but the details of a communal creative effort within the present state of affairs.

Might these pauses lead you to a pause over our present state of affairs, where *our* really means your particular Anthropocene concern? Might the pausing twirl you back to your own whatting to bring forth a sense that it would be as well to get a grip on the more fundamental wind-up: to make it our business to find out what's what? To arrive at a diagramming that would points us effectively towards a screw-up of our present state of affairs?

Ant Hope

Philip McShane

As I begin this second essay that invites further musings on a good and positive Anthropocene, I puzzle over whether I should add a complex note that might help towards a better reading. My answer to myself now, and to you, is, NO. Let that noting, noticing, be postponed till the next little essay: then it could prompt a push for much better reading, indeed cycles of such better readings.

The artists who are involved in the Anthropocene movement—naming them would be a distraction—will know what I mean on a preliminary level. The first notes of a symphony are heard differently, when one has listened to it before. Indeed—and this is a deeply important pointing—if one is a good and endlessly-improving listener, one may even accelerate in one's appreciation on each repeat. I recall a remark of von Karajan in his seventies about a Beethoven symphony he was to conduct. He had worked through a summer on what was to be, as I remember, his second-last recording of those symphonies and, when someone who knew of his intended autumn conducting remarked "will you not be bored?" he answered, "For me, it will be a new symphony."

But please, let not my musings and hidden references and von Karajan tale distract you from a simple sympathetic beginning to my story about a story to be written. Should we begin again: Once upon a time there was a global colony of ants who had settled patterns of harvesting and storing and busyness, patterns that, without their notice, blocked their fullest real business. One day a little group of them met a grasshopper who danced around and played the fiddle and the fool but who needed food. Was the grace hoper's fiddling just fiddle-faddle?

Many of you have gotten the reference here, at least to Aesop, if not to James Joyce's "Let us consider the casus, my der little cousis (husstenhasstencaffincoffintussemtossemdamandamnacosaghcusaghhobixhatou xpeswchbechoscashlcarcarcaract) of the Ondt and the Gracehoper."[1] Let us slip on now past this paragraph, to which we will take a little read-turn in the next essay, and move on here in seeming plain speaking.

Once upon a time, then, and the time is now: and I have added an "e" to the name of the first essay. It shifts "hop" to "hope" in this second essay of a later

---

[1] *Finnegans Wake*, 414, lines 18-21; most of the four lines are taken up by the single long (100 letters) bracketed word, omitted here from the brackets in the quotation.

now. But was there not a nudge towards hope already lurking in the first essay? That we were only at a beginning of our time? That we might bring culture uphill as Archimedes did water through a bit of screwing round?

The screwing around that I have in mind at the moment reminds me of the "mirror" poise, simply represented by Michael Jackson's song about looking in the mirror to change the world, a poise more complexly present in various masters and mistresses of modern dance. One observes one's moves in feedback fashion, and the observing is beneficially discomforting. I could extend this creative discomfort to all the arts, but now my interest is in the literary arts and those within that world of language—add the world of journalism. I want you to join me in screwing around with language, not in the way James Joyce did, but in a simpler and shockingly deeper way. Are you up for a little screwy shock?

There are different levels of shock, and I recall an elementary one that I used in a little book, my *Wealth of Self and Wealth of Nations: Self-Axis of the Great Ascent*. When I wrote it in the early 1970s, I had no idea that I was tackling the problem of getting us out of the negative Anthropocene in its early and then truncated forms. But that points to the third of these essays, so let now rather me give the elementary instance of screwy feedback, one that I clearly recall a publisher telling me "You can't do that!" Here is what was found offensive: "You may well at this stage read on, thus showing to yourself that like so many others you have suffered the standard failure in education, the failure to learn how to read."[2]

This is a very elementary feedback, jolting some readers to correct their failure in that text at a very elementary level. There is such a non-articulate elementary feedback in the Anthropocene movement as it emerged. Thus, the nudging of a broad human failure leads people like Edward Burtynsky to puzzle, in his personal website[3]—linked to the Anthropocene Project that he shares with Jennifer Baichwal and Nicholas de Pencier—about a shift to what he calls *A Good Anthropocene*. The puzzle, the puzzling, is expressed in a question mark, "?" Let us listen to the music of his puzzling:

But in the face of inevitable human influence on the Earth, what does #AGoodAnthropocene mean? It means a move towards sustainable solutions. It means actively reducing environmental pollution and

---

[2] McShane, *Wealth of Self and Wealth of Nations: Self-Axis of the Great Ascent* (2nd ed.), 17–18.

[3] Edward Burtynsky, "A Good Anthropocene," https://www.edwardburtynsky.com/news-hub/2018/9/5/a-good-anthropocene.

degradation. It means a global, collective effort to live consciously and responsibly. The list goes on.

In #AGoodAnthropocene I, along with Jennifer Baichwal and Nicholas de Pencier, am trying to exercise and employ creativity and passion to address the challenges our planet faces. While much of the work throughout my career has focused on the often irreversible influence our species has had on the Earth, the idea has never been to vilify or blame. Rather, this art has sought to align these landscapes with a more active and engaged awareness around our own implication in these processes. Likewise, *The Anthropocene Project* as a whole seeks to both inspire and expand understanding. The shifting of consciousness is the beginning of change.

You have read, listened, perhaps ingested Burtynsky's poise and concern. Might I, even in such success, repeat discomfortingly my nudging from page 20 of *Wealth of Self and Wealth of Nations: Self-Axis of the Great Ascent*, quoted above? "You may well at this stage read on, thus showing to yourself that like so many others you have suffered the standard failure in education, the failure to learn how to read."

In my book, the problem was obvious: an explicit puzzle was skipped and felt at some level to be skipped. In reading the piece from Edward Burtynsky there is most likely no sense of skipping. Burtynsky is pointing to a problem stirred up by feedback: Climate change affects us and, in some places, unbreathable air is screaming at us. But now I nudge you to hear the scream that lurks in the text, or should I not say lurks in the reader of the text and in the passionate art to which the text refers?

I pause here over the possibilities of identifying this scream effectively, but decide, yes, a halt till we venture forward in the next essay might poise you in some lung lunge to air, to care for, a lurking discontent. Indeed, might Edward and Jennifer and Nicholas take a screwy turn to find what lurks? An incomplete repetition might help here: might Edward and Jennifer and Nicholas take a turn to find what lurks. No print mistake here, in the dropped question-mark: **what** lurks. Listen to **what** lurking at the end of Burtynsky's short comment: "*The Anthropocene Project* as a whole seeks to both inspire and expand understanding." What seeks? What seeks. It seeks understanding, indeed an effective understanding.

But might it not seek better if **what** knew what's **what**?

The quotation from Burtynsky ends with the claim: "the shifting of consciousness is the beginning of change." But his claim is in a fog of his not-knowing what's **what**. The threesome of #AGoodAnthropocene are in a cloud of unknowing that they share with their opponents, their challengers. "In #AGoodAnthropocene I, along with Jennifer Baichwal and Nicholas de Pencier, am trying to exercise and employ creativity and passion to address the challenges our planet faces." So, we have, in this threesome, good will in each trying to exercise something whose nature is unknown. Further, the unknowing is not the unknowing of the early negative Anthropocene, but of the axial truncated negative Anthropocene where the unknowing has become a tradition of education and manipulation, a brutal serfdom. #AGoodAnthropocene is just not going to work before we run out of planetary sanity. All the movements of Anthropocene art and activities shrink to being an Ant Hop.

Ant Hopper

Philip McShane

There is no need to be mysterious about the nominal meaning of my title. ***Ant*** refers to us in our present destructive global state. ***Hopper*** has various meanings to be found in dictionaries but—I quote Webster's New World Dictionary, 3rd entry—my meaning here is "a box, tank, or other container, often funnel-shaped, from which the contents can be emptied slowly and evenly."

You have probably seen such structures, ways of controlling a flow into a horizontal pileup. Now think of the problem we left off with at the end of the first essay in this series: getting a water zone to flow up into some sort of higher level of water. Here Archimedes' water problem is replaced by a problem of lifting global culture. So, we arrive at the end of the second essay, or more precisely at the middle of the quotation from Edward Burtynsky: the problem is to find "a global, collective effort to live consciously and responsibly." We are back at the problem of deliberation, and indeed—now think in terms of feedback—at getting, by deliberation, some decent view of and ventures in deliberation. But we need a strange Ant Hop here, a Catch-22 Ant screwing. Did Socrates have some sense of this when he threw a grasshopper slogan of leisure into Athens' Ant dens: "Know yourself," as he tried, e.g., to raise the serf's knowing in a square-doubling exercise? If you don't know knowing, how can you know yourself? But what is knowing? Yes: what is knowing.

If you are with me here, even in some slim way, then you are probably not an existential member of the truncated negative Anthropocene. Such a membership requires education of the type that darkens the essay by Dennis Brown, the first page of which I reproduce on the next page. In the previous essay I mentioned, at the beginning, complexities that could be treated, noting that I would turn to them in this essay. But I do so only as an aside, as something that you would notice were you to travel through various websites on the Anthropocene focusing on those, of a literary turn, that are interested in it. Frankly, I do not see the search for sniffs of the coming positive Anthropocene from that area. This, in spite of my own identification of such sniffs in, e.g., Ezra Pound, James Joyce, Marcel Proust. Later I'll come to aspects of this identification.

But if you are curious about this problem, you might well venture into Brown's article and his learned weaving of ondt and gracehoper round tensions between Wyndham Lewis and James Joyce. Where is it all going, you may ask: the where the going is getting you, if you are foolish enough to thus venture, is into the silly overreach of all these folks into the Space and Time of Einstein (see the final paragraph of Brown's essay). It is a going that is part and container-parcel of the sad way of contemporary sophistications of putterings around knowing, putterings that are hilariously and sickeningly truncated.

James Joyce's Fable of the Ondt and the Gracehoper: 'Othering', Critical Leader-Worship and Scapegoating

Dennis Brown

'The Gracehoper was always jigging ajog, hoppy on akkant of his joyicity.'[1] This quotation from Joyce's *Finnegans Wake* has been used as an epigraph for the first chapter of a book on spirituality and prayer.[2] Such an instance serves to demonstrate how Modernism's 'Safety Pun Factory'[3] has percolated into the cultural sphere in diverse ways and usually, as here, in brief evocative passages. In fact, the main passage about the Gracehoper is short enough to stand as an anthology piece, and has become one of the best-known sections of Joyce's extraordinary text. In his 'joyicity', the Gracehoper clearly operates as a fond simulacrum of the writer himself, and as a kind of assertion of the attractions of his way of being:

> the sillybilly of a Gracehoper had jingled through a jungle of love and debts and jangled through a jumble of life in doubts afterworse, wetting with the bimblebeaks, drikking with nautonects, bilking with durrydunglecks and horing after ladybirdies. ... (416)

This amounts to what Norman Mailer has called 'advertisements for myself',[4] an affirmation of precisely those aspects of Joyce's lifestyle which worried close friends and supporters such as Harriet Shaw Weaver: his fecklessness, money-squandering, alcoholic binges, sponging, sexual prurience and so forth. The Gracehoper is nothing if not self-regarding: turning Aesop's binary value-system upside down, Joyce inherently endorses the Grasshopper's sunny *jouissance*. Arguably, this was his way of playing the 'stage Irishman', a Brendan Behan *avant le fait*, and quite like the Joyce caricature in Tom Stoppard's *Travesties*: 'to the ra, the ra, the ra, the ra'. (415) However, implicit within the Gracehoping persona is the telling reposte Stoppard uses to bring down the curtain on his first act: '"And what did you do in the Great War?" "I wrote *Ulysses*. ... What did you do?"'[5]

---

[1] James Joyce, *Finnegans Wake* (1939; London: Faber and Faber: 1982), 414.

[2] Alan Ecclestone, *Yes to God* (London: Darton, Longman & Todd, 1984, 1975), 7.

[3] "Your wholesale Safety Pun Factory" – a phrase in Harriet Shaw Weaver's letter of February 4, 1927, quoted in Richard Ellmann, *James Joyce*, rev. ed. (Oxford: Oxford University Press, 1983), 590.

[4] Normal Miller, Advertisements for Myself (1959).

[5] Tom Stoppard, *Travesties* (London: Faber and Faber, 1975), 65.

From the first words of *Finnegans Wake*—'riverrun, past Eve and Adam's—it is clear that the book is centrally about language itself: "'The only thing that interests me is style', as Joyce remarked to his brother Stanislaus.[6] The elaboration of the Gracehoper's sexual proclivities demonstrates Joyce's method at its most suggestive – an exemplification of Julia Kristev"s 'Desire in Language':[7]

> he was always making ungraceful overtures to Floh and Luse and Bienie and Vespatilla to play pupa-pupa and policy-pulicy and langtennas and pushpygyddyum and to commence insects with him, there mouthparts to his orifice and his gambills to there airy processes, even if only in chaste, ameng the everlistings, behold a waspering pot. He would of curse melissciously, by his fore feelhers, flexors, contractors, depressors and extensors, lamely, harry me, marry me, bury me.... (414)

The reader who is prepared to adjust expectations and respond to this kind of intercommunication has to be patiently attentive. The voice is that of a children's story-teller ('The Gracehoper was always jigging ajog'), and many of the word-compounds are of an infant, rather than adult, sphere: 'Vespatilla', 'pupa-puap', 'pushygyddyum', and so forth.[8] However, other words evidence an adult salaciousness: 'insects' (if read as incest), 'orefice', 'fore feelhers', 'extensors'. There is also a 'transgressively' prurient quality in the language. But stronger even than this is the 'crossword puzzle' challenge of ambivalent reference: "Floh', 'langtennas', 'melissciously', and in particular (later) 'schoppinhour'. Joyce is inviting us to play an unusual language-game: but it is a game very much on his own high-cultural terms. The language is devised to elicit admiration, awe and a sense of initiation into élite hyper-consciousness.[9]

I pause here over Brown's last footnote in his essay, note 54. It is attached to the claim, "like psychoanalysis-in-practice, responsible criticism is about 'attention and interpretation.'"[10] Recall our venture mentioned at the beginning: a

---

[6] Quoted by Richard Ellmann, *James Joyce*, 697.

[7] Julia Kristeva, *Desire in Language: A Semiotic Approach to Literature and Art*, ed. Leon S. Roudiez, tran. Thomas Gora, Alice Jardine and Leon S. Roudiez (Oxford: Basil Blackwell, 1981). The book is about language in general, of course; but there is considerable mention of Joyce, including references to his "joyous and insane, incestuous plunge summed up in Molly's jouissance or paternal baby talk in *Finnegans Wake*," 151.

[8] Originally meaning 'unable to speak', of course.

[9] Dennis Brown, "James Joyce's Fable of the Ondty and the Gracehoper: 'Othering', Critical Leader-Worship and Scapegoating," n.d., 32, http://www.wyndhamlewis.org/images/WLA/2000/wla-2000-brown.pdf.

[10] W. R. Bion, *Attention and Interpretation: A Scientific Approach to Insight in Psychoanalysis in Groups*, (London: Maresfield Reprints, 1984).

search for a Hopper, a 'Container' that would be effective in our reach for and later cultivation of the positive Anthropocene, which by now perhaps you have some sense of as being grounded in asking 'what's what?' Chapter 7 of Bion's book is titled, "Container and Contained." The chapter is written within the high walls of truncated scholarship, with—check if you have the energy—nary a question mark in sight, much less a show of interest in what's what. You can google Bion and his output if you wish, but I see no point in identifying his nonsensical container here.

Move from the commentators to the writers: are these better poised? Well, let us start with Brown's view of Joyce, as a lead into listening to the musical timethumbs of Joyce. A paragraph on page 34 of Brown's article begins with the sentence "Overall, Joyce has constructed a sly and amusing 'container' within which to rehearse differences between selfhood and the other." Near the paragraph's end there is the supplementing comment, "Joyce knows exactly what he is about and has no need for psychoanalytic 'diagnosis' (C. G. Jung felt he was 'a latent schizoid'). In Kleinian terms, one might say that he has absorbed Lewis's oppositionality and transformed it reparationally into a literary game of 'Who's Who.'"

To the contrary, Joyce's massive creative innovations were carried out and contained, despite his adventures with Aquinas, in a cultural warp of truncation that left its marks and quarks on his who's who ventures. Certainly, he shows the twists towards subject that bent him to read the book of himself, but the bent is blocked.

Yet, within the language bump that blossomed from *Ulysses* into *Finnegans Wake* there was his growing appreciation of Ant or Fin "again" in the mood of Vico's *Scienza Nuova*.

Here I sense a problem that eventually resulted in the cut-off brevity of the fourth of these Ant essays. There is, in Joyce's limited effort, a pale shadow of the linguistic feedback needed for self-reading. There is the less shadowy raising of the issue of phyletic feedback. I could go on here to sketch these developments towards what I would call *Pastmodernism*. But how sketchy, how sadly?

I recall my title of 2002: *PastKeynes Pastmodern Economics: A Fresh Pragmatism*. I began that book with a quotation from Isaac Babel's *The Beginning*, about the path to be travelled by the honest revolutionary talked of by Gorky to Babel: 'A writer's path, dear dreamer, is strewn with nails, mostly of the larger sort'. Isaac Babel, as James E. Fallen flags in his title, *Isaac Babel: Master of the Short Story*, was magnificent in brevity, as Joyce was. I greeted that Joycean *Dubliners'* brevity in my Poundian

Cantowers: so *Cantower 8* of my long series of *Cantowers* is titled "Slopes: An Encounter," weaved round Joyce's short story, "The Encounter." It deals with the Ant Slopes onto which we must venture to reach the Grasshopper slopes in our hearts, grassmolecules nudging our feet towards feeling "The field"[11] of Being. In that 24-page essay I gave decent indications of a sloping, a substructure of the rescue-dynamics of crippled humants. Yes, I could go round again, screwing up, in a transposed Archimedean container, the waters of culture in my own minding: but what of your minding? I am calling loudly over the field. I am calling grass hopping and collecting containers in my fiddling over nails when I repeat here my initial quotation in that Cantower, from the end of James Joyce's "An Encounter." Might you do better than seize me by the ankles? "I went up the slope calmly, but my heart was beating quickly with fear that he would seize me by the ankles. When I reached the top of the slope I turned around and without looking at him, called loudly across the field."

That 24 pages ends with a paragraph that begins with *might*. The might is what and what's what. It is the might of the ant's "container" pictured properly as a box on page 36 of *Wealth of Self and Wealth of Nations: Self-Axis of the Great Ascent* (2nd edition). It is the box that is to contain the positive screw-up of the ants in later cultures, the eightfold whirl of my *Futurology Express*. On the following page of *Wealth of Self* I recall Tennessee Williams: "We are all condemned to solitary confinement within our own skins," but one should add Sinead O'Connor's voice in a remembrance of the future: "I have a universe inside me / Where I can go and spirit guides me / There I can ask oh any question."[12]

Can? Might? The final paragraph goes on—through Entropy—to end with what? With what. With: ?

Might we come, in this next century, to envisage the end-times of energy and entropy with sufficient heuristic light to integrate into imaginative syntheses, resonant with particular cultures, the patterns of the masts and masks and masterstrokes of Grace, the final frontier, that lace anastomotically into the geography of our galactic inner and outer wonderlands?

---

[11] Lonergan, CWL 18, 199.

[12] "The Healing Room," from the CD Faith and Courage.

Ant Racks

Philip McShane

My dictionary gives a wide variety of meanings for *rack*: you might find it enlightening to muse over them, find how many of them tie in with my present effort to talk of containers, frameworks, tortures, brain-racks, whatever. But perhaps you might be nudged better by musing over the strange paralleling that lurks in what my title echoes. Think, then, of this standard description:

> **Anthrax** is a serious infectious disease caused by gram-positive, rod-shaped bacteria known as *Bacillus anthracis*. Although it is rare, people can get sick with **anthrax** if they come in contact with infected animals or contaminated animal products.

The Ant Racks that I am thinking of, pointing to, are not rare. They are global: they are in the air we breathe and in the heirs we breed: we are inconstant contact with the contaminated products that are meshed into fiscal malice. More properly and precisely, they are in the massaged neurodynamics of all of us. We are, in fact, not ants, but grasshoppers. But the axial negative Anthropocene has bent our neurochemistry to servitude, an industrious devolution. The serious infectious diseasing of humanity possesses even those who profess to be gracehopers: so there are those who push for *The Coming Convergence of World Religions* as if that convergence was an ANTayeclimax. I have pointed in the opposite direction in the spring volume of *Divyadaan: Journal of Philosophy and Education* 30/1 (2019). But the racks and the rackets do not allow the pointing to be effectively noticed.

As I moved through the third of these essays it became sadly clear that I was perhaps pointing in vain, indeed edging towards repeating a vain pointing in which I paralleled in the venture of my *Cantowers*, at greater length than Ezra Pound, Pound's *Cantos*. So I decided to halt my pointing in this fourth essay, waiting to see whether people caught in the racks and the rackets of Anthropocene criticisms might shake their neuroheads a little bit towards "a resolute and effective intervention in this historical process."[1] Might the July 2019 conferring in Vancouver, regarding the positive Anthropocene, seed some shake up, shedding, shredding, shrinkage, of the Ant Racks?

---

[1] Lonergan, CWL 18, 306.

VI. The Need for Cyclic Thinking

Crecycling *Sustainability*

Philip McShane

My three colleagues and I conveniently split the task of reflecting on the book *Sustainability and Peaceful Coexistence*, so that each of us pause over one of the four parts. Obviously, my zone was part III, and it would seem that I have completed the task by reflecting on chapters 6 and 7. But that is not so: I see our problem—the problem of the four of us and of the growing crowd concerned with the present threatened world, as one of recycling. *Recycling* is a familiar word, and indeed an activity that we do in our kitchens and malls, with varieties of waste. We recycle all the time of course, in the sense of following routines, one breakfast after another, one war after another.

One war after another? A jarring suggestion, unless we are fatalists. And if we are of the group concerned with what I call the *positive Anthropocene*, war is not on the list of inevitable recyclings. Nor is it on the list of the people who wrote *Sustainability and Peaceful Coexistence*. So let me indulge in a neologism—these are abundant in the said volume—and suggest to you that we begin to think of crecycling. Crecycling is simply a compact name for creative recycling, but I am leading towards us giving it a special meaning, an overarching meaning for the reach of all groups concerned with the condition of our cosmos. The special meaning comes from crecycling, in its very new sense, my reach, in the conclusion to the previous essay, into the work of Arne Næss.

I do not wish to bother you with a reach into Næss's work, or into his recycling of Gandhi's reflections. So, I will be minimalist here in simply jumping off from the quotation from him with which I concluded my reflections on chapter 6 "'It's Getting Better and Better, Worse and Worse, Faster and Faster': The Human Animal in the Anthropocene": "Applied to humans, the complexity-not-complication principle favours division of labour, *not fragmentation of labour*."[1]

The jump-off is illustrated by considering our common journalistic knowledge of modern, even ancient, goings-on in science and technology. A problem emerges as given or givens: data. Tribal minders mess with it towards a breath of fresh looking-at: early times were pragmatic, reaching to make a wheel rather than define a circle, but even a wheel had a look-site in mind named concept. Is, was, the look-site and look-sight right? Well, it worked. Think, now,

---

[1] Arne Næss, "The Shallow and the Deep, Long-range Ecology Movement. A Summary," *Inquiry: An Interdisciplinary Journal of Philosophy* 16, no. 1–4 (January 1, 1973), 97, https://doi.org/10.1080/00201747308601682.

89

how we may proceed, here and hereafter? I am inviting you, indeed, to think now in terms of a concluding hint of my Ant text: more in terms of weighing how to lift water than dip crowns in it.

Cut back from the pragmatics—such a cut back was a mistake in history that I do not wish to deal with here—and you find a standard view of science that is our present ethos. A theory is verified in instantiating data. Furthermore, within that ethos nowadays there is an expectation of division of labor into three zones: complexity-not-complication: each member of the three groups implicitly identified, knowing precisely what they are at and about. There are the watchers, the thinkers, and the checkers. Note, for instance, that the watchers are not without thought: they are, at their best, in the full realm produced by previous thinkers. Think of the watchers of data flow in the research into fundamental particles. And so on, with the three groups.

I am here, you may have noticed, winding freshly round the last three paragraphs of the previous little essay. But notice the 'pushes forward', crecycling moves, where the word *crecycling* will continue to lack my suggestion, in the conclusion of this essay, of a *Praxisweltanschauung* for our troubled times. How might we imagine now, simplistically, that full suggestion, in a manner that helps us puts the meaning of *Praxisweltanschauung* into a genetic sequencing, like a growing flower? Perhaps imagine a spiraling brought about by creative recycling? Here I can lift us further forward, crecyclingly, by appealing to a text mentioned at the end of my previous reflection, *Essay in Fundamental Sociology*. The words echo what I have been saying above.

> But what is progress? It is a matter of intellect. Intellect is understanding of sensible data. It is the guiding form, statistically effective, of human action transforming the sensible data of life. Finally, it is a fresh intellectual synthesis understanding the new situation created by the old intellectual form and providing a statistically effective form for the next cycle of human action that will bring forth in reality the incompleteness of the later act of intellect by setting it new problems.[2]

Here I wish to be introductory, crecycling both Arne Næss's short text and this little text from Bernard Lonergan. The issue for both, and for especially now for the Anthropocene movement people, is "the transformation of the sensible data of life." Crethink that data. It is—you find this shocking, odd perhaps?—a

---

[2] Bernard Lonergan, "Essay in Fundamental Sociology," in *Lonergan's Early Economic Research*, ed. Michael Shute (Toronto: University of Toronto Press, 2010), 20.

lean-forward data, even if you are only screen-watching particle tracks. How much more so if the data is a sunflower? Further, the watcher of today watches through the lens that is today's guiding form. For the particle watcher it is the present lean-forward *Standard Model*, as it is called. The climate watchers, too, are in an ethos of today's heat-shifts and anxieties and are—certainly the watchers of this volume are—leaning forward towards a Standard Model of watching for our times. "Not on my watch" is a familiar phrase. The watching, normatively, is caring. But in this first Næssian "division of labour" its care leans forward within whatever Model is available—glorious in physics, shabby in economics—to detect, inspect, the vibrant data for nudges towards progress. We can comfortably call this detecting *Research*. Muse briefly, perhaps, about the existential fact that the nudges of Research can be either to the watching poise's careful attention or to the poise's glimpsing of seeding a suggestion regarding progress.

Identified nudging of "the guiding form" badly needs, in our future times, a divide between it and Research and what may be called here, for pedagogical reasons, *Storycheck*. That naming of a third division of labour throws handily into our minding the naming of our second division, our present concern, as *Storyform*. But before commenting on that division and its odd name let me note the mood that is to be present in the full "statistically effective form for the next cycle of human action." The mood is caught in the slogan, "This is worth crecycling!" The slogan is lurking in much of the volume on *Sustainability and Peaceful Coexistence*. But now we are in the ballpark of Næss and Lonergan, and you and me, crecycling the volume to find "a statistically effective form for the next cycle of human action."

In chapter 6 of *Sustainability*, LeVasseur finds "that Sale's thesis has very strong exploratory power."[3] Sale's thesis is a Storyform of 25,000 years of, e.g., ice and weaponry. What this second division of labour is to work with, and seek to add to, is a storyform of storyforms that somehow holds together the best of all such efforts as that of Sale. Is this a crazy crecycling of mine? On the contrary, the nudge sits there—ho ho worth crecycling—in the storyform of mathematics or in the storyform of the popular TV series *House* (2004–2012). A contemporary mathematician works in the story of stories, in the sequence of achievement identified in a good history: think of the recent break-through in Fermat's Last Theorem. House is a more available take-off towards a glimpse of storyform in its reduplicative hospital fullness. The team of experts bring a mastery, even with special regional masteries, to bear on the data of a patient's illness.

---

[3] *Sustainability*, 89.

Our patient is a sick globe: we need to aim for the highest poise possible of mastery, even if our present putterings seem reducible to the recent (December 2018) Polish climate-change conference laughter at USA's promotion of coal. The point is simple: we need to bring to our cherishing of data-nudges a grip not only on a standard model but on the storied sequence of such models. Newton and Maxwell haunt our hunt for the Higgs-thing. Our hunt can be called *Interpretation*, but its value depends on the hunters being weaponized by the full story of hunting for interpretations.

Suppose the hunt is successful: there emerges a new story-angle, a new standard model or at least a decent crepiece of the present one. But the question arises: Is it, in a very full sense, **right on**? Again, a simple illustration helps: the med-lab is nudged to think out a new pattern of chemicals, but do they, did they, will they, cure? So, musing with such and other illustrations, you can begin to think of *Storycheck*. A tricky business of literally facing artifacts and datafacts with freshened storyforms. I'm sure part of your problem with all this so far is that the lean-forward leans you towards thinking of the future as a concern of data-sifting, story-forming, and story-checking. Poise again over the question, "did they cure?" Were you not leaned, inclined, to think of the move forwards to sampling of new drugs?

So: we pause to savor that these three divisions of the labor of salvaging the human story are past-oriented, even though our bent and the data-bent is forward. The question of the caring watchers in any of these three zones is: What has happened to us? But we hope, as did Todd LeVasseur with Sale's story-venture, that, in its freshening of our story, there are seeds of salvage.

How are we doing so far?

I pose thus my delightfully tricky question, one about what we are doing here, but primarily because it is one that gives rise to a fourth division of labour, one that is not clear in the writings of Næss or the passage from Lonergan or indeed in the work of Kirkpatrick Sale. Yet, is it not there, a full caring of these writers? They write, certainly, caring for their text: but they write caring for salvaging. They are assembling: in Sale's case reaching way back; in Lonergan's case, hovering over the nationalisms of the 1930s; in Næss's case over a past study of ecology. They brood over what they have assembled seeking to find, yes, hope from a freshened basis, tweaked like a decent equation surrounding the Higgs particle, or quite eye-shocked up-stares, as by the little original quantum equation of Schrödinger.

So, their struggle is with the achievements of three divisions of labour, but it has its own poise of crecycling: précising the present overall and overarching best basis of storyform-making that has so far emerged from the anthropic scene. Certainly these people, pre-*Praxisweltanschauung*, pre-positive Anthropocene, are caught in present muddled searchings, but think of the benefit of having the assembled work of the prior zones being handed to them, salvaging them, as Næss would have it, from inefficient fragmentations of labour?

Perhaps think of us, here now, with the simpler version of the question, How are we doing? We have an assembly in the book *Sustainability*. We are, perhaps neither listed as authors nor mentioned in the massive collective of the bibliographies. Have we not a leisured focused advantage over the full collective? How are we doing? Are we beginning to see that the massive problem the book lays before us screams for a division of labour, and might we not leap, like the water in Archimedes' Screw, to a new level of seeing and seizing the effective connectedness of a newly watered watcher-culture? The leap occurred in and on the level of discernment that adds a twist on whatever discernments occurred on the way up through data-collecting, story-refining, story-checking. Its twisted poise has us asking: How are we doing in asking "How are we doing?" We wish all the spontaneous discernment that haunts the previous three divisions of labour to cease to be spontaneous and become self-luminous. Were we actually in that team of fourth-division carers we would weave our efforts—but now I am into a fantasy of the future—around our own luminous discernments to a strange but brighter water-level, a more-than-Schrödinger-shift of his thinking about "What is Life?"[4] that would lift the ocean of our minding to a geohistorical ethos of discerning discernings of discernings.

That final sentence is a dense opaque expression[5] of a massive cultural lift of minding that needs to occur in this century, while on the side we battle in little ways the dominant arrogance and stupidity and lack of caring fantasy of the world's leaders. Perhaps, sadly, the tides and airs we face in the mid-century will indeed call out Deep Green Resistance, but there must be those who are wakened "to providing a statistically effective form for the next cycle of human action that will bring forth in reality" a tangled weave of Green and Greed that allow our heirs to breathe for another crecycling millennium.

---

[4] Erwin Schrödinger, *What Is Life? The Physical Aspect of the Living Cell*, (Trinity College, Dublin lecture series 1943), (Cambridge: Cambridge University Press, 1944).

[5] We are heading now for a certain lift-off in our essay, but here I find it worthwhile to point to my densest expression of future global salvaging. It appears in Philip McShane, "Method in Theology: From $[1 + 1/n]^{nx}$ to $\{M (W_3{}^{\theta\Phi T})\}^4$," *Journal of Macrodynamic Analysis* 10 (2018), 105–35.

We are poised here, in our How—might I suggest, **H**ome **O**f **W**onder—asking, at the fifth and sixth step of the step-diagram as the end of "Crecyling *Insight*."[6] It is the turn to the future. Might it be the same group as assembled the past, turn now to a guidance system that, yes, brings to my curious mind Hedy Lamarr's guidance system creativity rejected by the U.S. military in the Second World War.[7] What the U.S. military at the time lacked was fantasy meshed with respect for women.

But back to the question of group-identity. My answer, of course, is No, No, Nanette: this is a T for two, a task for two groups. Here I seriously recommend a return to, or a first read, of my few Ant essays. It takes a massive psychological burst out of the superego grip of the negative Anthropocene to mediate the daft optimism of the initial stages of the positive Anthropocene. There is a sense in which the image of excellence in relay hand-on is at its most challenging in the shift of this Task for Two. The front runner is facing forward in a new way, intent on going forward with a grasping hand inviting the baton, but now magically in a new race. Then, THEN,[8] on she goes. Or, in the case of my allegory, on he goes, for he is Jimmy Smith of *No No Nanette*. Google it: but, in any case, here you have a Google description of the beginning of the first Act of the 1971 production:

> Jimmy Smith, a millionaire Bible publisher, is married to the overly frugal Sue. Jimmy thus has plenty of disposable income, and, because he likes to use his money to make people happy, he has secretly become the (platonic) benefactor of three beautiful women: Betty from Boston, Winnie from Washington, and Flora from San Francisco.

Betty, and Winnie and Flora are to be, in musical, three Hurricanes, the mirror image of the three groups that cared for Storycheck, Storyform, and Dataflow that whirl the past towards a comprehensive wholesomeness in relation

---

[6] See p. 107 below. The same diagram appears without notes on p. 118.

[7] I parallel Hedy Lamarr's efforts with those of Lonergan in Philip McShane, "A Heady Folly," in *The Allure of the Compelling Genius of History: Teaching Young Humans Humanity and Hope* (Vancouver: Axial Publishing, 2015), 53–63.

[8] Savor my madness here. The title of *Cantower* 5 of 2002 is "Metaphysics THEN" Philip McShane, "Cantower V: Metaphysics THEN," accessed June 2, 2022, http://www.philipmcshane.org/cantowers.). It begins with one of the last poems of Beckett: "go where never before / no sooner there than there always / no matter where never before / no sooner there than there always." When the water carrier of my final diagram hits a mature pace, it will be "there always." *Cantower* 4 had focused on revolutionary ladies, as odd as Hedy Lamarr.

to the Whole Human Gaia Story. What they need is a Platonic benefactor, who has a cre-up-to-date Screw-lift *Praxisweltanschauung* on the master's ancient appeal:

> Unless philosophers become kings in the cities or those whom we now call kings and rulers philosophize truly and adequately and there is a conjunction of political power and philosophy . . . there can be no cessation of evil . . . for cities, nor, I think, for the human race.[9]

My interest here, as I end this first part of the essay, is getting your serious attention on Flora, on Dataflowing forward, on Dataflowers forwarding, on those who are to be street-effective in causing a flow forward that, like Archimedes screw, will raise the water level, but now of culture and care. A nice accident, that: the Latin, *Floruit*, and the English Flora-wit. We have a few decades to get our wits into seriously-effective mode: "for there to be a resolute and effective intervention in this historical process."[10] I have had us climbing imaginatively but not realistically, effectively, towards the flowering of a first glimpse of the full intervention, and that must be an element in the effort of 2020–2050. The key elements are those pointed to, in scattered fashion in *Sustainability and Peaceful Coexistence* and in my next essay, while the urgent issue is to focus on a stumbling effective ordering of those random pointings. The ordering is of only airy use without the strange fresh effects being seeded: we must thus reach what are called *the masses*.[11] That is the topic of my next essay, possibly titled: "The Masses and

---

[9] Plato's *Republic*, Book V, 473c11–d6.

[10] Lonergan, CWL 18, 306.

[11] There is an obvious reference here to Ortega's *The Revolt of the Masses*. But I would note that Ortega's notion of the masses was quite complex. Chapters 6 and 8 of the book are directly on the topic, but also chapter 12 on "The Barbarism of Specialization." Saul Bellow, in his Foreword to the translation, neatly sums up Ortega and also the problem of the changes in the meaning of mass man since Ortega's time. "Ortega when he speaks of the mass man does not refer to the proletariat: he does not mean us to think of any social class whatever. To him the mass man is an altogether new human type. Lawyers in the courtroom, judges on the bench, surgeons bending over anaesthetized patients, international bankers, men of science, millionaires … differ in no important respect from TV repair men, clerks in Army-Navy stores, municipal fire-inspectors, or bartenders. It is Ortega's view that we in the West live under a dictatorship of the commonplace" (José Ortega y Gasset, *The Revolt of the Masses*, trans. Anthony Kerrigan, University of Notre Dame, 1985). As my steps below intimate bluntly, most of scholarship, a great deal of science, and all of politics, are done by mass men and women inviting themselves and us to settle into such a commonplace. The problem of that talk is raised in profound doctrinal fashion in the first section of *Insight*

Sustainability." Still, here I wish to continue our imaginative climbing in the mode of Næss but with an acknowledgment of Bernard Lonergan's parallel mode. Finally, I would note that I have already, in this concluding of the first section, added nudging footnoting to my mode of communication. There is the simple melody of my text, a beginner's sing-along of *Für Elise*, florawise—or perhaps of a hymn-version of the *Ode to Joy*; there are the footnote chordings beneath, pointing to a strange Bruckner 8th symphony climb to the end of Beethoven's 9th.[12]

---

chapter 17. The next section offers a discomforting large and remote solution that is enlarged further by the pointing of my final diagram: "Here's a step we can't afford to miss!"

[12] The climb to the choral is steep enough, but I add the reference to Bruckner's 8th because it has been symbolic for me of the climb to effective functional scientific collaboration: a five-note echo trickling in at the beginning of the second movement and finally taking over the symphony: so, we trickle in at, we hope, the beginning of the second movement of the Anthropocene.

Philip McShane

> Take a little one-step, two-step, three-step.
> Come a little closer, please,
> like a rose that blows in every breeze.
> Take a little one-step, two-step, three-step,
> then a little dip, like this.
> Here's a step we can't afford to miss!

Do not be distracted by my weave of this lyric from the 1971 ending of *No No Nanette* into my plea. Is the weave even vaguely relevant? I recall weaving in references of the more recent X-Factor shows into my invitation to get to grips with "The Well of Loneliness" in you.[1] The weave is vastly relevant in both cases, as I try to intimate. But let me come to the main pointing here. That pointing is towards your closer foundational dancing and singing. There is the foundational dance of chapter 6 of *Insight*, "Common Sense as Subject"; there is the second one two three steps of chapter 7, "Common Sense as Object", and then a little dip like towards an X-factor gathering: "here's a step we can't afford to miss!"

(i) Take a little one-step,

Our step is in and around the first sentence of chapter 6 of *Insight*. "The illustrative basis of our study must now be broadened." Bracketed by four words on each side we have "our study." My grandson Matthew, just over three at the time of writing, can talk of "study," though he means a room, grandma's study, not an activity. Or does he? In his second year he rolled up his t-shirt displayingly and used the word *boobies*: he likely had in mind his mother's delightful brown milkshakes, now his little brother's joy. What of such meanings in the distant human future? What of schooling meanings in later millennia? How, then, are we to know our meaning for *our study*, our growing symphony of meaning? Your symphony of meaning pivots on an early age achievement of Helen Keller's leap at age seven.[2] But does it include the meaning of that meaning? So, we meet our

---

[1] "The Well of Loneliness" is the title of chapter 19 of *The Allure of the Compelling Genius of History* (Vancouver: Axial Publishing, 2015). The X-Factor shows are introduced there, in relevant fashion, on page 225.

[2] This is my step-challenge here. It is not an easy step. Pointers to it are given in Philip McShane, *A Brief History of Tongue: From Big Bang to Coloured Wholes* (Halifax, NS: Axial

first little one-step. The illustrative basis is broadened: "most strikingly illustrated by the story of Helen Keller's discovery."[3]

But what was that discovery? Was it not a discovery of what? But it was not a discovery of what's what, no more than it was for Matthew at two. You had that discovery, or you would be now just g-aping at this page. But have you ventured into a month, parallel to Helen's month of March 1887, with or without an Annie Sullivan, to discover this core piece of what's what?

My little one-step about a little one-step is a discomforting nudge about the climb to foundations, to a sufficient luminosity to get us out of the present slum of minding. I nudge the seeding of "the illustrative basis" to a self-redemptive task, a personal stance on what got the ape into the Anthropocene. "It is with the basis of this much more personal stance that the fifth functional specialty, foundations, is concerned."[4]

Are you concerned, or are you content to grope along with lurking pretentiousness in the commonsense dark?

(ii) two-step,

We move forward, with Matthew and Helen, to Flora, and the massive problem of florawit, the full global cultural supporting lean-forward of "the incessant What? and Why? of childhood."[5] "They flower only if we are willing, or constrained, to learn how to learn."[6] The foundational cycler needs an ongoing crecycling of learning to learn. That cycler, in our times, has to detect the spectrum of bluffs that have been established, with brutal antkill destructiveness in the long history of education. But now I am talking to you about the challenge of entering a cycle of bitten grasshopes.

We need a leap to new talk and new words.

Pause now with me over two quite different pages. I quote a decent piece of the first that identifies the poise of talk of educators and leaders the world over, then I shock you with Newton bites.

Press, 1998), 31–37. The absence of this scientific insight into a common basic human insight takes the heart out of psychology: leaves it happily, as it is now, in complexifications of initial meanings, regularly reductionist.

[3] *Method in Theology*, (1972), 70; CWL 14, 68.

[4] Ibid, 267; CWL 14, 250.

[5] *Insight*, CWL 3, 197.

[6] *Insight*, CWL 3, 197.

By classicism I mean the fruit of an unsuccessful education in which, first of all, there is no real grasp of theory of any kind—mathematical, scientific, philosophic, or theological. Theory is proposed and studied, but in the subject there is no serious differentiation of conscious; all we get as a theory are the broader simplifications offered by a professor to introduce or round of a lecture or a course, or the product of *haute vulgarization.* But he is never bitten by theory; he has no apprehension, no understanding, for example, of the fact that Newton spent weeks in his room in which he barely bothered looking at his food, while he was working out the theory of universal gravitation.[7]

This passage, I have no doubt, has been read solemnly by scholars—indeed even perhaps by you—without the temptation occurring to try a bite of Newton. Well, here goes: read and weep, or skip past this peril of great price.[8]

---

[7] Bernard Lonergan, "Time and Meaning," in *Philosophical and Theological Papers 1958-1964,* ed. Robert C. Croken, Frederick E. Crowe, and Robert M. Doran, Collected Works of Bernard Lonergan 6 (University of Toronto Press, 1996), 155.

[8] The peril is that we otherwise continue (see note 11 on page 95.) to foist on the future "the arrogance of omnicompetent common sense" (Lonergan, CWL 17, 370). The analysis quoted below this in the text is from my pre-notes of a first-year university course on mathematical physics given in 1959-60. I am quoting page 19 of the notes on Statics, reproduced as no. 7 of my website articles (http://www.philipmcshane.org/website-articles). These were pre-notes: the lectures were, so to speak, off the cuff, but in this sophistication of language, taken for granted and shared by the good students. We spoke explanatorily. Where, one may ask, is language to go to rescue us from global arrogance? Linguistic feedback, like this nudging, is a feeble but essential start.

Given Kepler's Laws:
(i) ellipses;
(ii) equal areas are swept out in equal time;
(iii) T^2 (prop to) cube of mean distance from S.

Find what forces?

From (i): $\ell/r = 1 + e \cos \theta$ ($e = .0167$ for earth)

From (ii): area $(\frac{1}{2}) r^2 d\theta$ in time dt: rate $(\frac{1}{2})r^2 (d\theta/dt) = $ const. $(\frac{1}{2})h$ say.

If we assume F_r (inward) per unit mass and F_θ p.u.m.

we get the eqts of motion:

$d^2r/dt^2 - r(d\theta/dt)^2 = -F_r$

$(1/r)d/dt (r^2 d\theta/dt) = F_\theta$

but from (ii) $r^2 d\theta/dt = $ const. h.

therefore $(d/dt)(r^2 d\theta/dt) = 0$; therefore $F_\theta = 0$.

This simplifies maths, if we write $u = (1/r)$;

then $d\theta/dt = hu^2$.

We eliminate r from the other eqt. of motion:

$r = 1/u$

$dr/dt = -(1/u^2)(du/d\theta)(d\theta/dt) = -h(du/d\theta)$

$d^2r/dt^2 = -h(d^2u/d\theta^2)(d\theta/dt) = -h^2u^2 (d^2u/d\theta^2)$

so $d^2r/dt^2 - r(d\theta/dt)^2 = -h^2u^2 (d^2u/d\theta^2) - h^2u^3$

Eqt. of motion in r becomes

$-h^2u^2 (d^2u/d\theta^2) - h^2u^3 = -F_r$

We want to determine F_r:

And we have u in terms of θ:

$u = (1/\ell)(1 + e \cos \theta)$

$du/d\theta = (-e/\ell)\sin \theta$

$du^2/d\theta^2 = (-e/\ell)\cos \theta$

so

$F_r = (du^2/d\theta^2)h^2u^2 + h^2u^3$

$= h^2u^2[(-e/\ell)\cos \theta + (1/\ell)(1 + e \cos \theta)] = (h^2u^2)/\ell$

Therefore $F_r = (h^2/\ell)(1/r^2)$

SO: we get to Newton's inverse square law.

(iii) Three-step,

"Out of the plasticity and exuberance of childhood through the discipline and play of education there gradually is formed the character,"[9] but the plasticity can be

---

[9] *Insight*, CWL 3, 212. I would note here a massive topic that connects with the content of note 6 on page 98. I quote the first paragraph of the Aristotelian text, *Magna Moralia*: "Since our purpose is to speak about matters to do with character, we must first inquire of what character is a branch. To speak concisely, then, it would seem to be a branch of nothing else than statecraft." The central flaw in the topic's treatment in that text is the topic of the conclusion of Philip McShane, "Finding an Effective Economist: A Central Theological Challenge," *Divyadaan: Journal of Philosophy & Education* 30, no. 1 (2019), 97–128.

toys that rust psychically, and the education can be, and indeed is in our time, a parallel glossy scummy slum.

The three-step is the finding of therapy, and my toes must what-move to find the need in my grass-seed boxed in ceding in sweet daily daze to a pervasive, a global, scamoutants.[10] How can the grass-hoper weave a me-in-my-corner field? Some HOW one must field the need. The censor and the superego, aided by the slum lords, toss the consentient ant back and forth in the quiet rhythms of a "narrow orbit, for each is free yet together swept in a swirling mass down the cataract of life to the serene pool of a green churchyard."[11]

If there is any green left.

"The analyst, then, is needed."[12]

But there is no ontic[13] analyst up to the job: nor—(ix) "then a little dip, like this"—is there a phyletic analyst unless we begin to settle, settle-up, for effective hope in the newrowglow of an unknown X coming your way in the (X!) step below.

(iv) Come a little closer please,

"There is a further and deeper aspect to the matter [of intellectual development]."[14]

The deeper aspect lurks there in chapter 6 of *Insight*. Indeed, it lurks in the reading of the first word of the first chapter: if you really get In. But chapter 15 helps to come a little closer if you please. There is the ape and then there is the muddling apemanant scammed out by the limp twaddle of a slim whatting inventiveness. Would Betty and Winnie be better, winners, on their own, with perhaps an island man on a Friday? The habitat could then be organized by a solo,

---

[10] Beyond the obviousness of "scam-out-ants" there is the hiddenness of the reference to the scotoma spectrum touched on in *Insight* CWL 3, 213ff. Roughly, we might talk of the neurofirming of the massive axial sick superego talked of in note 11 on page 95.

[11] Bernard Lonergan, *Shorter Papers*, ed. Robert C. Croken, Robert M. Doran, and H. Daniel Monsour, Collected Works of Bernard Lonergan 20 (Toronto: University of Toronto Press, 2007), 78.

[12] *Insight*, CWL 3, 225.

[13] See note 11 on page 95 and note 10 above: the analyst is in the axial ballpark. The broader problem is described in Lonergan, *Phenomenology and Logic*, *CWL* 18, chapters 13 and 14.

[14] *Insight*, CWL 3, 498. Recall that *Insight's* chapter 6 begins with "Common Sense as Intellectual," 196.

an aria in the symphony of nature, bowing to the ineffable. But organizing a flight of arias needs a flow of Archimedean screwed-up symphonies in nature to raise the solo voice. No woman is an island. Still, nature is a silent communing even to an apewomanant: a field of dreams, telling her that the habitat is what's soil and sonatina, guiding her actions by "referring them, not as an animal in a habitat, but as an intelligent being to the intelligible context of some universal order that is or is to be."[15] The referring, in its neurobreak of antegotic superego, if it is effective, comes closer but not with ease please. One can write of "the music of the spheres!"[16] at the end of a long life and in the conclusion of a dark play, a Mozart unfinished Requiem. So, Betty and Winnie and, yes, ladies like Nadia Boulanger and Georg Eliot, can grow to let the field "dominate a whole way of life."[17]

(v) like a rose that blows in every breeze.

What is this rose, this X-Factor that can breeze from the world's stage? What is this *Dance of the Rose* that ends with Nijinsky's window leap in our rise? Is there not a moment in your rose garden, even when torn with thorns? Times when (a Kavanagh phrase) "the millstone has become a star"?

> To what indeed shall I liken
> the world and human life?
> Ah, the shadow of the moon,
> When it touches in the dewdrop
> The beak of the waterfowl.[18]

These cries lift up the whatter in the water foul of truncated times, making the whatter one with the universe, so that

> it shares its dynamic resilience and expectancy. As emergent probability, it ever rises above past achievement. As genetic process, it develops generic potentiality to its specific perfection. As dialectic, it overcomes evil both by meeting it with good and by using it to reinforce the good. But good will wills the order of the universe, and so it wills with that order's dynamic joy and zeal.[19]

---

[15] CWL 3, 498.

[16] Shakespeare, *Pericles*, V, i, 229.

[17] *Insight*, CWL 3, 498.

[18] Dogen (1200-1253), Waka on Impermanence (#61-J).

[19] *Insight*, CWL 3, 722, conclusion.

(vi) Take a little one-step,

Stepping into *Insight*'s chapter 7 was seemingly—until you were led to take five, iv, iii, ii, i, steps—a modest business for your common sense, a business of tuning into a curious view of common sense. "The apparently modest view of common sense is to understand things in their relations to us" begins the chapter with the nudge to that lead of mine in its second word. Apparently? A parenting of you as individual. Recall (ii). "Complete free play to intelligent inquiry" is psychically blocked, and you continue to be led ant articulateness and live in a world of talking heads. But the leading away from inquiry is supplemented by an egobent that "will not grant serious consideration to its further relevant questions."[20] Its? Both the ego and the bend. This is best seen in others who are bent to slip past creative puzzling.

One of my favorite directives to my class, for twenty years, of young ladies was to air their reports of dates, especially on the what-to-do question. "What are we going to do tonight?" "The usual." And we have so many ways of disguising the usual as fresh moves.

"Relevant questions" ends the section of *Insight* 7 on "Individual Bias." Did you move immediately to read the next words, "General Bias"? And how, prey, are you moving here? Might you not pause and take a little step in the dark?

(vii) two-step,

Groups bind feelingfully, whether local gangs or giant nations, tennis twos or choral tens. Group bias winds that bind of feelings into a shrinkage of care, a swelling of group care. Four dense pages of chapter 7 of *Insight* skims over its twists and turns, but here I hit home to all shrunken groups by poising over an irritating one. "I do not think there is any need to flog a whole row of dead horses; a flick at a particularly nauseating one is enough."[21] The group I am sadly thinking of is the group to which Lonergan originally passed his baton and the school that festers his genius at present. You know who you are, but likely enough few of you are reading this. The group originated out of gentle presentations of Lonergan, a stranger taken in. "I was a stranger and you did not take me in." (*Matthew* 25: 43). Should I put Lonergan's strangeness in the past tense?

> Was my proposal utopian? It asked merely for creativity, for an interdisciplinary theory that at first will be denounced as absurd, then

---

[20] *Insight*, CWL 3, 247.
[21] Lonergan, *For a New Political Economy*, CWL 21, 36.

will be admitted to be true but obvious and insignificant, and perhaps finally be regarded as so important that its adversaries will claim that they themselves discovered it.[22]

(viii) three-step,

To take the three-step away from "General Bias" is to be, I dare to say, quite inhuman. The negative Anthropocene lives and breeds in that bias. For millennia initial meanings, bubbled spontaneously from primitive cultures in a rich web of feelings and associations, and then yielded sickly to elementary self-discovery mistaken by Jaspers as the axial period of humanity, thus establishing truncated initial meaning in that lonely zone. Initial meanings? "An accurate statement on initial meanings would be much more complex,"[23] indeed it will take more than a few centuries of the positive Anthropocene to effectively articulate it as dead. And it would take more than a few hundred words to skimpily intimate it. But heavens, then, is there really no difference between the various wonderful talking heads, like Fareed Zakaria and Arnold Toynbee, except the extent of their competence in contextualizing magnificently initial meanings?[24] And what of Wolf Blitzer's solemn situation room? Etc etc.[25] So nominalism, regularly splashing in numbers, non-comprehending technological competence in developed lower sciences and messy upper sciences rules, in sweeping and sweet brutality, our nations and notions.

(ix) then a little dip, like this.

The little dip, of course, is the mind-wind round the mess that crimeshes battered Apeanthows into antics and academic disciplines sniffing and sniffling round the stench of decay and death on this old globe. There is a dance of death in limbs and words that pretends hope in prose and cons. "The ooze of abnormality"[26] secretes the pretentiousness into gatherings of art and science in old, trapped

---

[22] Bernard J.F. Lonergan, "Healing and Creating in History," in *A Third Collection*, ed. Frederick E. Crowe (New York: Paulist Press, 1985), 108.; CWL 16, 103.

[23] CWL 3, 568, n.5.

[24] We are, of course, in the problem pointed to in notes 8, 10, and 13.

[25] See, chapter 12 of Philip McShane, *Profit: The Stupid View of President Donald Trump*, (Vancouver: Axial Publishing, 2016): "*The Situation Room: The Stupid View of Wolf Blitzer*." You may be shocked by my directness: it is an essential component of the future collaborative science of dialectic: see the final lines of page 250 in *Method in Theology* (1972), CWL 14, 235 (end of Section 5).

[26] *Insight*, CWL 3, 262.

styles. The negative Anthropocene turns directors and actors towards "become stagehands. The setting is magnificent; the lighting superb; the costumes gorgeous; but there is no play."[27] Need I go on? The volume on *Sustainability* finds no serious place in lobbied business of glocal leadership: we are at Sixes and Sevens.[28] What is this century screaming for? A cosmopolis. "What is cosmopolis? Like every other object of human intelligence, it is I the first instance an *X*, what is to be known when one understands."[29] Might it wipe out, in millennia to come, our silly nationalisms?

> What is necessary is a cosmopolis that is neither class not state, that stands above all their claims, that cuts them down to size, that is founded on the native detachment and disinterestedness of every intelligence, that commands man's first allegiance, that implements itself primarily through that allegiance, that is too universal to be bribed, too impalpable to be forced, too effective to be ignored.[30]

(X!) Here's a step we can't afford to miss!

Not to miss the stepladder needs a wild dervishness. Google the dance that goes with the 1971 *No, No, Nanette* lyrics with which we danced along mind-folly. It is a pretty thing, a pretty plain thing, almost a solitary Celtic jigging. What **what** needs in the present mess is indeed something dervish yet communal, quite a leap beyond the addition of those lyric of 1971 to the original musical of 1925. Perhaps, the leap is helped by an image of a leap in quite another zone. We have the problem of someHOW getting the whatter up a hill of culture. Back we go, then, in imagination, to Archimedes. No no Notyet, is there a need for you to get to grips with HOW he did it. Just admire his finished product with a sense parallel to the primitive reaction to the revisioning of simple Celtic dancing that was Michael Flatley's *Lord of the Dance* of 1996: a heart-beat line-up of drumming feet. Here you have Archimedes' structured water-dance:

---

[27] *Insight*, CWL 3, 262.

[28] Would Robert Plant's "Sixes and Sevens" lyrics help? Would anything help us to rise to another arrangement for our house of cards, our home-going? "Sundown, another busy day watching the time fly / Old ground standing in the way / And I don't know why / So here I am making changes / Alterations to my house of cards / But I don't hold new arrangements /Am I at home, am I at home, am I, am I alright?"

[29] *Insight*, CWL 3, 263.

[30] *Insight*, CWL 3, 263.

Think of all the Molly-cures of Water tuned to dance uphill. Now think of Bernard Lonergan doing a Jimmy Job for the Molly-cures of Whatters: Tom, Dick and Harry, Betty, Wendie and Flora. You might think of his early choreography, his writing in 1935 about an apparently simple native circle dance of global intelligence: let's repeat the dance-notes here, thinking of the Zulu proverb, "The isisusa wedding dance is always appreciated by being repeated."[31]

> But what is progress? It is a matter of intellect. Intellect is understanding of sensible data. It is the guiding form, statistically effective, of human action transforming the sensible data of life. Finally, it is a fresh intellectual synthesis understanding the new situation created by the old intellectual form and providing a statistically effective form for the next cycle of human action that will bring forth in reality the incompleteness of the later act of intellect by setting it new problems.[32]

In 1971 he finished his effort to talk out the details of his new dance.[33] The talk fell on deaf toes. But that is another story. Might you just and justly pause

---

[31] I quote here from the first major work ever written in Zulu, referenced and commented on in note 10 (p. 166) of my *Lack in the Beingstalk* (Vancouver: Axial Publishing, 2006), one of my previous efforts to point to the massive cultural transition symbolized in my final diagram of this essay.

[32] See note 2 on page 90.

[33] What I have attempted here can be considered to be a fresh presentation of chapter 5, "Functional Specialties," with two differences: I refrain from developing the final three specialties; I do not give his "grounds for the division" (*Method in Theology*, 1972, 133; CWL 14, 127). My grounding here is in the emergence in history of the need, such as was seen by Næss, for a division of labour. Leaving out the final three specialties seemed wise. I have elaborated on them abundantly elsewhere: for example, secularly, in

now, see, oddly and shabbily, the twinkling toes of Arne Næss wending and winding his way to the full ten-step diagrammed here, oddly and shabbily?

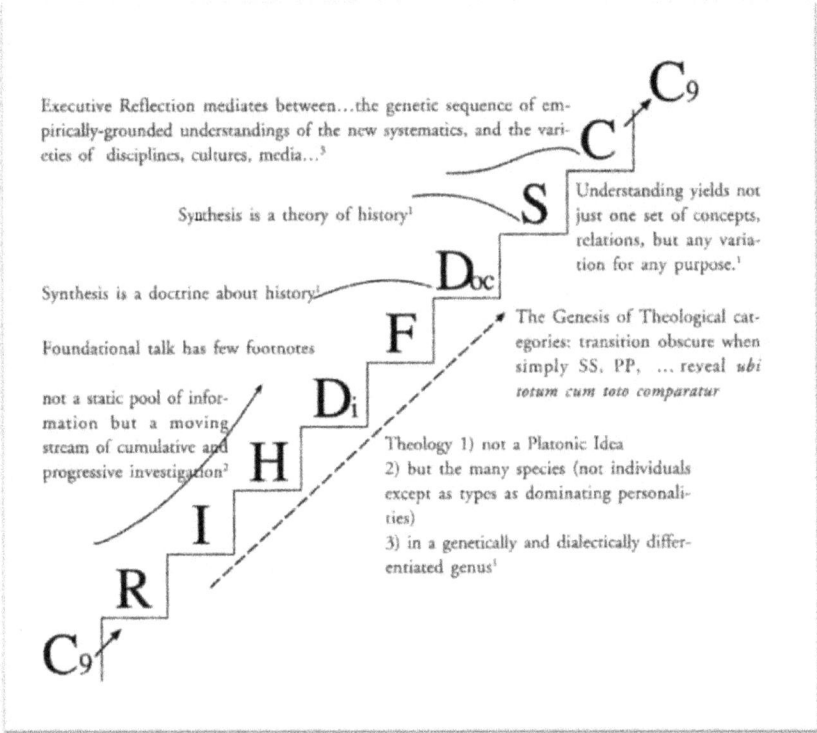

Executive Reflection mediates between…the genetic sequence of empirically-grounded understandings of the new systematics, and the varieties of disciplines, cultures, media…[3]

C_9

C

Synthesis is a theory of history[1]

S

Understanding yields not just one set of concepts, relations, but any variation for any purpose.[1]

D_{oc}

Synthesis is a doctrine about history[1]

F

Foundational talk has few footnotes

The Genesis of Theological categories: transition obscure when simply SS, PP, … reveal *ubi totum cum toto comparatur*

not a static pool of information but a moving stream of cumulative and progressive investigation[2]

D_i

H

Theology 1) not a Platonic Idea
2) but the many species (not individuals except as types as dominating personalities)
3) in a genetically and dialectically differentiated genus[1]

I

R

C_9

1. Bernard Lonergan, from unpublished notes of the early sixties available in the Toronto Lonergan center, Batch B, 8, 6, V.

2. Bernard Lonergan, "Christology Today: Methodological Reflections," *A Third Collection*, ed. Frederick E. Crowe, S.J. (Mahwah, NJ, Paulist Press, 1985), 82.

3. Philip McShane, "Systematics, Communications, Actual Contexts," *Lonergan Workshop*, vol. 6, ed. Frederick Lawrence (Chicago, CA: Scholars Press, 1986), 151.

Futurology Express (Vancouver: Axial Publishing, 2013) and, in a Christian context, in *The Allure of the Compelling Genius of History* (Vancouver: Axial Publishing, 2015). The missing notes on the ten-step diagram are available on page 189 of *Allure*, or on the version of the diagram's first appearance (1990), in chapter 4 of *Process: Introducing Themselves to Young (Christian) Minders*. The latter book is available at http://www.philipmcshane.org/website-books. The key introductory work to the massive shift of the crippled masses talked of above in notes 8, 10, and 13 is *Wealth of Self and Wealth of Nations. Self-Axis of the Great Ascent* (2nd ed. 2020) (1st ed. 1975, available at: http://www.philipmcshane.org/published-books).

The Masses and Sustainability

Philip McShane

There is a definite sense in which the strange diagram with which I ended the Crecycling *Insight* is to be as familiar to the masses in the 22nd century as the diagram of the periodic table in chemistry became in the 20th century. There was a time, indeed when I first produced that diagram in 1989,[1] when I thought it might be thus a familiar thingy for 21st century people. Might you have a hand in rushing to reality my hope of 30 years ago? Helped on, of course, by our present crises of the globe.

And perhaps helped on by this follow-up essay, which, as originally planned, was to puzzle creatively with you over chapters 6 and 7 of the book, *Sustainability and Peaceful Coexistence*, but now invites you to puzzle over the whole book in a simple fashion, indeed in a fashion that relieves you of the task of reading the book. So, "puzzling with you" takes on a less sophisticated meaning. I am assuming only that you are interested and concerned about the global mess we are in. You belong to the concerned masses as I specified *the masses* in a footnote in Crecycling *Sustainability*, certainly worth repeating now below.[2]

---

[1] The diagram first appeared in chapter 4 of *Process: Introducing Themselves to Young (Christian) Minders* (1990). The bracketing of the word, *Christian*, indicates that the poise can be taken in secular fashion, with the simple omission of chapter 5. The book serves, with McShane, *Wealth of Self and Wealth of Nations: Self-Axis of the Great Ascent.*, as lead-ins to the task of rescrewing the Anthropocene.

[2] There is an obvious reference here to Ortega's *The Revolt of the Masses*. But I would note that Ortega's notion of the masses was quite complex. Chapters 6 and 8 of the book are directly on the topic, but also chapter 12 on "The Barbarism of Specialization." Saul Bellow, in his Foreword to the translation, neatly sums up Ortega and also the problem of the changes in the meaning of mass man since Ortega's time. "Ortega when he speaks of the mass man does not refer to the proletariat: he does not mean us to think of any social class whatever. To him the mass man is an altogether new human type. Lawyers in the courtroom, judges on the bench, surgeons bending over anaesthetized patients, international bankers, men of science, millionaires ... differ in no important respect from TV repair men, clerks in Army-Navy stores, municipal fire-inspectors, or bartenders. It is Ortega's view that we in the West live under a dictatorship of the commonplace." (Ortea y Gasset, *The Revolt of the Masses.*, ix) As my steps below intimate—with an obscurity that is to be lifted in the decades ahead—most of scholarship, a great deal of science, and all of politics, are done by mass men and women inviting themselves and us to settle into such a commonplace. The problem of

But now it is also useful to repeat the diagram and comment on it briefly in a way that helps the helping and self-helping. The repetition is less frightening in that it is not on the first page! On we go, then, as we try helping ourselves and eventually those who oppose us in this, yes, global war. So, think very simply of the fact that we need to get organized. Might I remind you here of Al Gore's efforts, and Al is one of many. But, as he is only too willing to admit, the efforts are as yet not effective. So, we need to get better organized, get into layers of collaborating organization: some layers, even, which—recall DEW—in the short run, may try meeting force with force. But let us now view the odd, apparently complex, diagram with simple practical eyes.

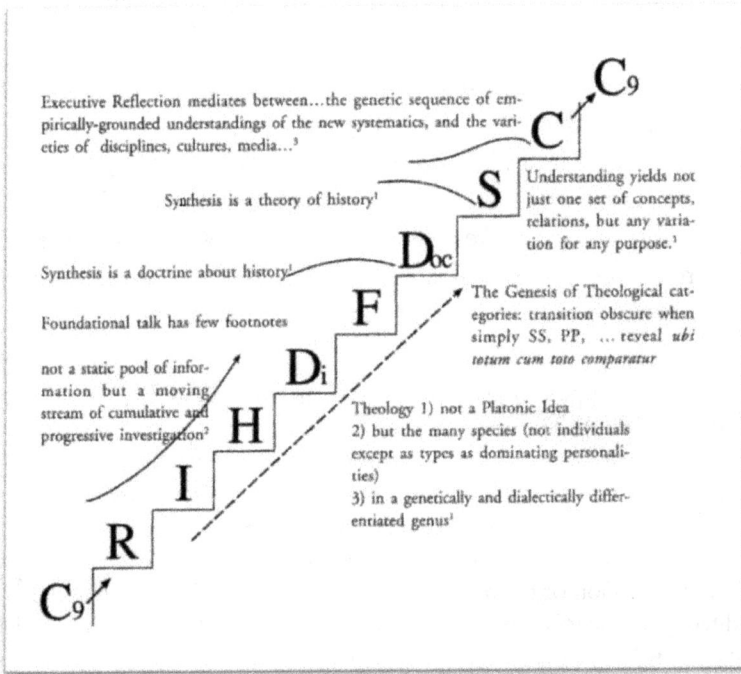

The bottom C_9 represents the problem zone, ourselves on this endangered planet. The 8 steps above that are symbolic of the real steps to be taken to get us all to be the top C_9, real steps that are to leave us masses better off. The real steps

that talk is raised in profound doctrinal fashion in the first section of *Insight* chapter 17. The next section (Crecycling *Insight*) there offers a discomforting large and remote solution that is enlarged further by the pointing of my final diagram: "Here's a step we can't afford to miss!"

are to emerge, much as the periodic table emerged after centuries of chemical messing.

But let us not get into the complexities of scientific messing. Think rather of "the need to organize" that nudges all of us regularly: I have to cook a dinner, coach a team, comment effectively on sustainability.

The final illustration of the previous paragraph obviously brings us back to our paradigmatic book, *Sustainability and Peaceful Coexistence*. So, we find that there is the obvious set of moves of searching out participants, understanding—in collaboration with them—their possible contributions, etc., etc. Might you follow up here on the etc., etc. and discover that, not too surprisingly, we can do a fairly simple climb up the steps to arrive at, yes, a sort-of top C_9, which is the final book, out there published, with some effects such as us here now taking it seriously.[3]

What I am asking now is that we take it seriously enough to face into a necessary crecycling of it. Necessary? The book has not worked, nor will it work, in bringing about "a resolute and effective intervention in this historical process"[4] that led us to the present crisis. Might we not invite the authors of the book *Sustainability* to consider this failure and this opportunity? Indeed, I do: but the immediate invitation is to you, and it requires crecycling of your own poise regarding the state humanity is in. That poise may well have been given a lift from the book, the spread of its references and bibliography, or it may simply be a general journalistic unease. But I do now ask for a pause over my diagram. And in that asking I shed my previous shifting intentions of helping you along, with, say, a sifting of the book's content that shows that all the analyses and suggestions made there can be ordered creatively by and into the ten-step diagram, with a creativity that is to be increasingly effective in this next millennium. That pausing and that sifting and that creativity is to be communal, related easily and encouragingly to the title I gave the movement it names, *OPA*, "Openers of the Positive Anthropocene." You are invited to become an **opal** by finding your position on one or perhaps two of the steps, and, whatever your positioning, to commit yourself to an effective attitude of reaching out locally to form **opacels**.

You need no expertise to participate in this effort. Sustainability, indeed, is the challenge of what I call *the masses*, standing against top-level possessiveness and

---

[3] Please pause over the cooking and the coaching illustrations: do you find some equivalent steps in stepping to success. My own coaching effort in the area, *Futurology Express* (Vancouver: Axial Publishing, 2013), begins with a family discovering that after 20 years, their holiday goings-on are a mess: splitting up the review of the mess opens doors to sane celebration.

[4] Lonergan, *CWL* 18, 306.

standing for the bottom-level dispossessed in a manner that is to green the globe, standing against the stale silliness of neoliberalism, of the voodoo of present economics and law, of mythologies of nation states.

VII. Structuring the Reach Towards the Future

Structuring the Reach towards the Future[†]

Philip McShane

> To see near things as comprehensively
> As if afar they took their point of sight,
> And distant things, as intimately deep
> As if they touched them. Let us strive for this.[1]

Prologue

The focus of this chapter is on the topic structuring or rather on re-structuring, since there are already structures and structurings and structuring-efforts among us regarding the reach that concerns us here. It seems to be vital as well as bright to structure my talk within the novel structure I have in mind, yet at the same time show its roots in present and past structure. That bright shift leads me to split this chapter in two. What is central and needing our attention is the topic of Part One: "An Effective Structuring of Peaceful Coexistence." Part Two, "Remembrance of Times Past and Future," deals, at least sketchily but in proleptic poise, with previous efforts to structure our existence, efforts that obviously range across a spectrum of what I loosely call *progress*.[2]

To be strict in my division I introduce, not in Part One, but in the Prologue, the key pedagogical nudge: the nudge given by Archimedes' invention of the apparatus that lifts water from a lower to a higher level.

[†]This chapter was first presented at *The 3rd Peaceful Existence Colloquium*, Helsinki, Finland, June 13–14, 2019.

[1] Elizabeth Barrett Browning, *Aurora Leigh*, Book 5, lines 189–192.

[2] I have used the words *effective, existence*, and *progress* here. Looseness of meaning is the name of the game here, though you may well think of our present meshed with the concern expressed by the existentialist movement of the twentieth century. I go on in the text to use the word *deliberation* and it shares the same looseness. Part Two will tackle the issue of the road to effective precisions of meaning. Finally, I would note that the center of our concern is a meaning of *effective* that is effective, shifting towards and beyond a Poisson statistics of success in our century to the Bell Curve of future millennia. Our present crisis is one of "effective shifting towards."

There is a range of lessons to be learned from that venture, that creative deliberation. The first lesson comes to the heart and hearts of our gathering: it is the lesson precisely of the need for creative deliberation and for luminosity regarding its own characteristics: to this I return in the Epilogue. The need for the activity is illustrated by a simple pause over Archimedes' leap of inventiveness; the need for characterizing it can be sniffed out slowly by simply pausing over the shabby attention "deliberation" has received in the intellectual traditions of humanity.[3]

Such a characterization cannot be a priori. It is an empirical business of attending to creative deliberations as they occur in more and more sophisticated forms precisely because of scientific progress and—may I use the phrase loosely for the moment—the engineering that blossoms from it.[4] The weeds of axial

---

[3] Here you meet a central problem of this chapter. Might I symbolize it by pointing to the gap between Aristotle (384–322 BC) and the crippled thinking of Peter Drucker (AD 1909–2005) both—note their dates—axial males? (On *axial*, see pages 134ff and notes 66, 67, 68, 69 and 71 below.) Suffice it to see that deliberation has not been seriously deliberated upon even if puttered skillfully round by a tradition that includes Aristotle, Nemesius, Damascene, and Aquinas. The paradigm represented by our first diagram of the screw has not had the deliberate attention it needs as symbolizing deliberation, and this cripples the movement for sustainability and peaceful coexistence. You may later follow my struggle here through notes 6, 17, 20, 22, 33, 45, 56. Then do the Hamlet (see the text around note 25) or Hal (see the text at note 59) thing, or the lady Sands thing of note 55: what-bore into your core.

[4] Engineering is to blossom only in the luminous boring of that core that asks, what-bright, "what might be, what might this be?" It requires a present subtle dismantling, a new mantle, a taking root of the long road of the new mantelling of the diagram on p. 118 below. See further, on mantelling, notes 16 and 49. The long road is the topic of the Epilogue to this chapter.

engineering, however, are a dominant reality.[5] Our gathering at *The 3rd Peaceful Existence Colloquium* is in the context of the present destructive sophistications, and, further still, that we are pressured by time. We have a Canadian television program, running since 2009, titled *Chopped*. The challenge there is to move from mess to meal in 30 minutes. I do not think that Archimedes was pressured in his deliberations, but we are. Have we thirty years to lift global living from present swamp waters to some sort of beginning of a sane waterworld? Let us pause, with this question, over an image of Archimedes' achievement.

How are we to raise the cultural waters so as to rescue and freshen the waters and bloodstreams of nature? Might I suggest extravagantly that we oppose the poise of Archimedes on science to Aristotle's poise?[6] But that is a teasing leap into and beyond Part 2 of the chapter. Let me just note that primitive humanity needed primitive science to work. It did not have a bourgeois interest in art or science for its own sake. A decent pause over how humanity got by in a pre-bourgeois non-ecumenic world[7] would gradually show that science is not a neat little academic three-step going to the moon but a ten-step collaborative global cherishing of earthlings and their cosmic home. That gradual showing is part of our larger task.

---

[5] The meaning of *axial* is a topic, too, of the Epilogue of this chapter. Perhaps it stirs the imagination a little to say that is it an evolutionary period that covers the Holocene age and the negative part of the Anthropocene.

[6] I pick up from note 3, Aristotle's "bourgeois" poise (see notes 43 and 69 below) that locked science into a three-fold way of verifying theory in data. Deliberating over Archimedes' deliberation is to push us towards a radical effective shift in our view of the disorientations of industrious humanity. On the bourgeois poise in the history of economics, see Geoff Mann, *In the Long Run We are All Dead: Keynesianism, Political Economy and Revolution* (New York: Verso, 2017). On the core of the road to economic science and sanity, see P. McShane, *Economics for Everyone: Das Jus Kapital*, 3rd edition (Vancouver: Axial Publishing, 2017).

[7] I am thinking now of that late volume, titled *The Ecumenic Age*, of Eric Voegelin's *Order and History* (The Collected Works of Eric Voegelin, Volume 17) (Columbia: University of Missouri Press, 2000). It ends in early China but does not empireism live on in megacorporations?

Part One: An Effective Structuring of Peaceful Coexistence

It seems best to begin with a diagram that encourages us, gives hope that we really can do this. Even though the beginnings of "this" are to involve messy skirmishing,[8] the strategy is to blossom through the next seven millennia into a global ethos of care.[9] So, here you are: a diagram of a cultural apparatus resembling Archimedes' screw for, in various good senses, screwing up civilization.

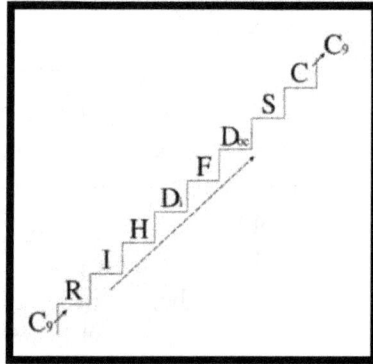

---

[8] Immediately I think of Todd LeVasseur's article "It's Getting Better and Better, Worse and Worse, Faster and Faster: The Human Animal in the Anthropocene," *Sustainability*, 87–100, and the varieties of resistance he presents, such as DGR (deep green resistance) and DEW (decisive ecological warfare). *Sustainability* also has essays that address the varieties of human ecological displacements that are likely to lead to more than skirmishing. I think of that idiot slogan in the low-grade film, *Independence Day*, "We will not go quietly into the night." So, yes, "worse and worse," but somehow the global plague poises an increasing number to take seriously the push of Bernard Lonergan, "insofar as there is to be a resolute and effective intervention in this historical process, one has to postulate that the existential gap must be closed" (*Phenomenology and Logic*, Collected Works of Bernard Lonergan 18, ed. Philip McShane [Toronto: University of Toronto Press, 2001], 306). Ruuska's *Reproduction Revisited: Capitalism, Higher Education and Ecological Crisis* (MayFly Books, 2018) [hereafter *Ruuska*] points vigorously to the gap. My effort here is to specify an effective dynamics of closing the gap. In the short term, we face agonies, such as those described in David Wallace-Wells book, *The Uninhabitable Earth* (London: Penguin, 2019), a book that will seed a little positive fright. And there is the larger positive that I draw attention to in note 32. I comment at length on LeVasseur's essay in chapter IV, "It's Getting Better and Better, Worse and Worse."

[9] See note 56, below. Add the fuller context of the Epilogue.

My presentational effort here is foundational, in a sense sought by Arne Næss forty years ago,[10] but it is steered by me here into a foundational pedagogy. To give a glimpse of that in another diagram helps us move forward pedagogically, even though it seems altogether too early for such a complexification. My conversation here is, in the letters of the above diagram, FC_9, and in the diagram below C_{59}. F points to Foundations, eventually to become a dominant social group, outwitting, in a cyclic collaborative dynamic, the remnants of the present "dominant fundamental group"[11] in its national and transnational varieties. C_9 points to the global community in its full historical concreteness, and if you like a cute image of what the reach in this conversation is, fancy the slave-built pyramids inverted to grant a munificent global microautonomy. On the next page, then, my third pedagogical diagram.

---

[10] In 1989, as I struggled in a sabbatical in Oxford to brood forward towards *Process: A Paideiad*, a detecting, leaning into India, of history's effort to educate us, I was astonished to find his detecting of a parallel structure of cosmic deliberation. My book was thus titled in its promise at the end of *Wealth of Self and Wealth of Nations* (1975; 2nd edition, 2021), but its final title is *Process: Introducing Themselves to Young (Christian) Minders* (Vancouver: Axial Publishing, forthcoming). These two books seed the present essay. For more on Næss, see note 58, below.

[11] I am referring here to Gramsci's view of guiding ethos. "The 'spontaneous' consent given by the great masses of the population to the general direction imposed on social life by the dominant fundamental group; this consent is 'historically' caused by the prestige (and consequent confidence) which the dominant group enjoys because of its position and function in the world of production." (*Selections from Prison Notebooks of Antonio Gramsci*, ed. Q. Hoare and G. N. Smith (New York: International Publishers, 1971, 12).

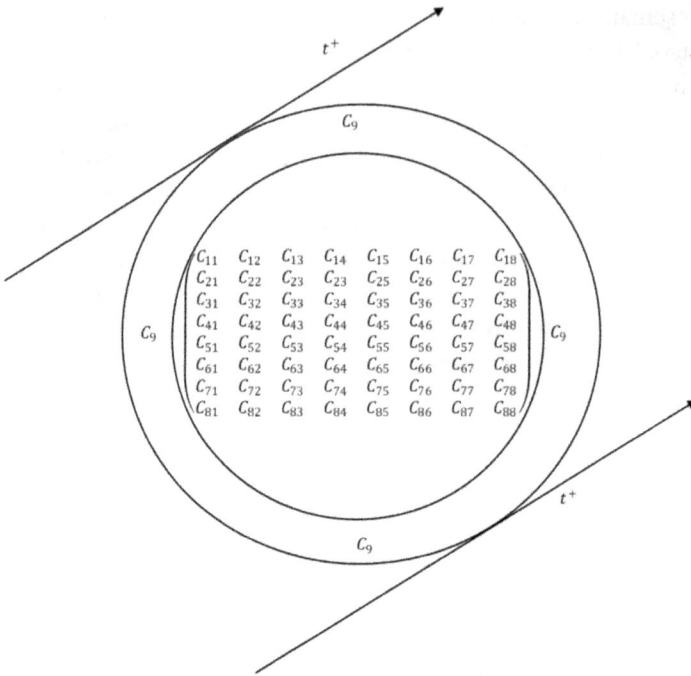

t^+

C_9

C_9

$$
\begin{matrix}
C_{11} & C_{12} & C_{13} & C_{14} & C_{15} & C_{16} & C_{17} & C_{18} \\
C_{21} & C_{22} & C_{23} & C_{23} & C_{25} & C_{26} & C_{27} & C_{28} \\
C_{31} & C_{32} & C_{33} & C_{34} & C_{35} & C_{36} & C_{37} & C_{38} \\
C_{41} & C_{42} & C_{43} & C_{44} & C_{45} & C_{46} & C_{47} & C_{48} \\
C_{51} & C_{52} & C_{53} & C_{54} & C_{55} & C_{56} & C_{57} & C_{58} \\
C_{61} & C_{62} & C_{63} & C_{64} & C_{65} & C_{66} & C_{67} & C_{68} \\
C_{71} & C_{72} & C_{73} & C_{74} & C_{75} & C_{76} & C_{77} & C_{78} \\
C_{81} & C_{82} & C_{83} & C_{84} & C_{85} & C_{86} & C_{87} & C_{88} \\
\end{matrix}
$$

C_9

C_9

t^+

Think of this spread as a new periodic table, but now the elements are human groups collaborating towards "redeeming time."[12] This diagram locates the previous step diagram as an inner community, an inner circle matrix, committed to a science of cosmic care which I have named *Futurology*.[13] But now we must ask together: What is this identification of the diagonal, the axis, C_{ii} of the full collaboration, C_{ij}, that is to deliberate cyclically, spirally, a vortex[14] of redeeming

---

[12] Shakespeare, *Henry IV, Part One*, I. ii. 210. The full line is "redeeming time when men least think I will," a suitable slogan for anyone who takes a stand on sustainability and peaceful coexistence. In the Epilogue I return to the full soliloquy of the prince (lines 188–210) which ends thus.

[13] *Futurology Express* (Vancouver: Axial Publishing, 2013) is the title of my recent popular presentation of this future poise. At present the train is in the ramshackled station.

[14] Yes, another image here, helpful, hopefilled. It comes from the eccentric Ezra Pound of a century ago. Pound wrote "if you clap a strong magnet beneath a plateful of iron filings, the energies of the magnet will proceed to organize form . . . the design in the magnetized iron filings expresses a confluence of energy" ("Affirmations, Vorticism," *The New Age*, XVI, 11, January 14, 1915, 277).

time from the mad destructive greed of the "civilized"[15] majority of the present global population?

Let us start now with vague identifications in, so to speak, the status quo of the academy, a shrunken business to be dismantled in these next centuries and mantled over millennia.[16] If you scan the various departments, you will find our eight steps, but not identified as such, and indeed in some departments some subset of these steps in proper sequence. Let us spell out the list, which is the diagonal list of the second diagram but diagonal in the left to right direction in the third diagram. Here we are:

Research; **I**nterpretation; **H**istory; **D**ialectic; **F**oundations; **D**octrines; **S**ystematics; **C**ommunications.

Note that I have suggested, by the break in listing, a division into two groups of four. The suggestion is of a past-orientated group and a future-orientated group, but keep the poise that belongs to all our present inquiry as you muse on this. The growing reality of the groups is to be such that, right from the first step of Research, this is a leaning forward enterprise. An image and name I like is that the spiral enterprise is to be ***A Leaning Tower of Able***.[17]

---

[15] Lurking in my essay there is a sense that we are no more civilized in this millennium than a sunflower is after a week's weed-pressed growth. Since Marx wrote, the masses have ascended into apparent financial comfort, but real enslavement. I nudge for a longer view and hope of "Arriving in Cosmopolis" (see note 55 below) and recall in that essay a note that I repeated in "The Masses and Sustainability," note 2, p. 109 above.

[16] *Ruuska*, 240ff gathers his poise in a final "Reproduction Revisited": I visit his poise in my final word here, my Epilogue. The word *mantelled* used above, perhaps a neologism, names the slow lifting forward that enlarges the context of Ruuska's poise to the objective of what I came to call, through writing this chapter in his presence, *PEM*, **P**rogress **E**ffectively **M**antelled. The mantelling is not management in any Drucker sense: it is to have a microautonomous luminousness of a topology of glocal situations symbolized by a complex sublation of F. M. Fisher's suggested imaging of history as a Markov matrix of statistics. See his "On the Analysis of History and the Interdependence of the Social Sciences," *Philosophy of Science*, vol. 27, no. 2 (1960), 147-158. My initial integration of this into my poise is in chapter 11, "Probability-Schedules of Emergence of Schemes," in *Randomness, Statistics and Emergence*, 2nd edition (Vancouver: Axial Publishing, 2021). The complex sublation is symbolized by {M $(W_3)^{\theta\Phi T}$}^{4}, where M is a spherical Markov structure of time-spreads of situations.

[17] Part of the struggle forward is a massive surge, in this century, of new imaging, new words that have a sting of explanatory linguistic feedback, to your molecular core, the language of a new "Æcornomics" (the title of my website series, available at

The "to be" is in our hands, is our challenge. At present there is no Tower, but a shambles. There will be a Tower, shaped up by our descendants of later millennia, the sooner if we shape up a seedbed now. We are the seedbed, we and those who gather with us, and the seeds are the twisted versions of the eight steps of the diagonal that are present in this pathetic state of axial humanity.[18]

Let us pause over this 8-fold mess of R, I, H, D_i, F, D_{oc}, S, C which at present is not a fold but a spread of parochial entrapments that holds all of us hostage.[19] Begin, perhaps with a touch of optimism, with the final three: D_{oc}, S, C.[20] Recognize in them the bent of university departments such as Commerce and Engineering and Political Studies. These departments work towards telling us, telling all of us, where we are going. The seeds of our future vortex threesome are named there, in varying versions, *Policying, Planning,* and *Executive Decisioning.* But in those and all other departments, there are the seeds of the first three in our list: R, I, H. Think of physics, with its threesome, data, tentative theory, verification, and then muse over literary studies as they include the threesome research, interpretation, history—critical or not. These threesomes are carried forward in

---

http://www.philipmcshane.org/ecornomics). The image referred to in the text is available in two forms on pages 94–95 of *The Allure of the Compelling Genius of History* (Vancouver: Axial Publishing, 2015). That book, while focused in a particular religious tradition, gives the mood of my effort here: we seek to detect the compelling genius of history, a way Gaia nurtures towards screwing up present inhumanity. See further, my five articles on the topic in *Divyadaan: Journal of Philosophy and Education*, vol. 30, no. 1 (2019). The volume is titled "Religious Faith Seeding the Positive Anthropocene Age."

[18] In these few notes I am deliberately tilting you forward in dark fantasy. I am recalling Herbert Marcuse's claim: "Without fantasy, all philosophic knowledge remains in the grip of the present or the past and severed from the future, which is the only link between philosophy and the real history of mankind" (*Negations: Essays in Critical Theory*, translated by Jeremy J. Shapiro, Boston, 1968, 155).

[19] Ruuska's book is a magnificent effort to reveal to us just how thorough the hostage situation has become: are we not dead meat, in a Patty He-arse (sic) situation? The question this chapter raises is, how that effort is to become effective over the next millennium. More of this in the Epilogue.

[20] We are here in the world of Drucker, who fathered management studies, and indeed contains seeds of our leaping beyond him. It is a world I decided not to enter in this short chapter. There are seeds of the leap, too, in Ruuska's stance as he pushes through the second half of his book. But we need a quite new view of the road forward that is to be taken by a massive global shift in our meaning of the care weaved into group carings of C_{66}, C_{77}, C_{88}.

their narrow ways, science for science sake and art for art's sake.[21] The inclusion of the words *tentative* and *critical* weave into both areas a bent towards the final pair in our list: D_i, F.[22] But the seeds of that pair are the heart of those ancient—oriental and occidental—departments of human concern: philosophy and religion. These seeds are generally battered within the narrow confined mindings of these academic disciplines but show sun-searching when they slip into artistry. Let us pause and cherish such a "moment in the garden."

> To what indeed shall I compare
> The world and human life?
> Ah, the shadow of the moon,
> When it touches in the dewdrop
> The beak of the waterfowl.[23]

I asked for a pause, a deliberate deliberative pause. To what, indeed, shall you and I compare the world and human life? Indeed, I place your pausing now in Hamlet's socks, poised in the first scene (lines 56–89) of the third act,

> Whether 'tis nobler in the mind to suffer,
> The slings and arrows of outrageous fortune,
> Or to take arms against the sea of troubles,
> And by opposing end them. (lines 57-60).

So, we arrive at a "perchance to dream" (65) that takes us towards considering an "undiscovered country" (79) that perhaps, would lead us into "the pale cast of thought" (85), thus "to grunt and sweat under a weary life" (77) so

---

[21] We are back at the broad problem of note 3, of a myth that would keep concrete care out of science and art, and indeed art out of concrete care.

[22] But you have noted that this chapter reaches out to, or into, you, precisely in these two phyletic group-zones. If I have not got you into some sort of dialectic poise, then you are reading this chapter as a capitalist academic! The chapter is a foundational communication that can be named C_{59}. Might it even get you to identify your work or your potential? But it should surely nudge you into that broad group C_{99} with some effective revolutionary poesy?

[23] The translation of a verse of Dogen, the Japanese Zen Master (1200–1253), from a Waka on Impermanence (#61-J), quoted in Heinrich Dumoulin, *Zen Buddhism: a History, Volume 2: Japan* (New York: Macmillan, 1990), 72. The original, scripted in English, is: *Yo no naka wa / nani ne tatoen / mizutori no / hashi furu tsuyu ni / yaadora tsukikage.* Translations vary. I replace in the text the word "liken" in the first line with "compare."

that, in a later millennium, Pericles, wife and daughter, would pause quite radiantly in the cosmos you hand to them: "The music of the spheres! List, my Marina."[24]

So we find our Hamlet moment and find, yes, the rub:[25] "perchance to dream. Ay, there's the rub." (65)

Indeed: for the evolutionary green is so far from smooth as be the hazardous mess with which we are familiar, a mess of which we catalogue catalogues in our gatherings and writings. Ay, there's the rub. To what shall we compare the world and human life?

We may step forward, like Shakespeare's *Pericles*, poised in a madness of hope of a distant human tree,

> … inflame'd desire in my breast
> To taste the fruit of yon celestial tree.
> Or die in the adventure.[26]

But now let us further pause, should I say prosaically?—a present ill[27]—and struggle towards some sense of the rub, by flexing our imagination towards the story of the rub.

It is a long story, seen as such by all of us, ice-aged in our psyche: but even so it is difficult to "remember the future,"[28] to see effectively the long way forward and the task of finding that way in the past. We can share the magnificent Hamlet dream-moment at the end of Toni Russka's book: "Capitalism is a historical

---

[24] Shakespeare, *Pericles* V. i. 229.

[25] The origin of the meaning of *rub* is an ancient game of bowls. A rub is some fault in the surface of the green that stops a bowl or diverts it from its intended direction. The word is recorded some few years before Shakespeare's time and is still in use.

[26] *Pericles*, I. i. 20-22.

[27] I treat of that ill, to some extent, in Philip McShane, "Aesthetic Loneliness and the Heart of Science," *Journal of Macrodynamic Analysis* 6 (2011), 51–84. See also Philip McShane, "*Æcornomics* 2: The Pedagogy of Trading Between Nations," 2019, http://www.philipmcshane.org/ecornomics.

[28] I am thinking of Proust's *Remembrance of Times Past*, which ends with Marcel "towering on giant stilts" of meaning. I think of us communally as moving towards such towering but lifting Marcel's sensibility into the world of integral self-appreciation. And I am also thinking of a chapter title, "Remembering the Future," a chapter on the West-of-Ireland playwright J. M. Synge in *Inventing Ireland: The Literature of the Modern Nation* by Declan Kiberd (London: Vintage, 1996). Synge and his seaweed culture! But a national feed of seaweed would cut out Irish cow-farts. The problem facing, us, however, is a see-feed to C_9.

structure. It *can* be replaced. It *must* be replaced. It *will* be replaced."[29] One can add to his climb there, from the vast catalogue of storytellers, a fuller telling of the "riverrun past Eve and Adam"[30] and sniff how we have soiled the riverruns of the world.[31] But how are we poised with *can, must, will?*

Strangely, capitalism, in the short run, is on our side.[32] The poise I write of certainly is to contribute to that twist of capitalism, but it is a poise of cosmic care that I now wish us to think about, a genetics of that care, a genetics of caring for that care. And that thinking is to carry us forward to a fresh musing over R, I, H.[33]

---

[29] *Ruuska*, 254.

[30] The beginning of James Joyce, *Finnegans Wake*.

[31] I am thinking of Tibet, the source of Asia's rivers. See Michael Buckley's *Meltdown in Tibet: China's Reckless Destruction of Ecosystems from the Highlands of Tibet to the Deltas of Asia* (Palgrave Macmillan, 2014). But you will find the evil near your own village. How might you mind it? So, I think also of the Czech river, the Vitala (wilt ahwa "wild water"), the Moldau (in German), and thus flow into Bedřich Smetana's *Má vlast* of the 1870's, and on into that river's weave into the world-rivers of James Joyce in *Finnegans Wake*, and so come now to recommend to you a stance walk-about as I did once—seven days is the summer of 2004—around the shores of my Dublin river: see my *Quodlibet* 8: "The Dialectic of My Town, *Ma Vlast.*"

[32] We need to cling to the long-term optimism to which I point in the Epilogue, but that clinging is to be an increasing psychic reality. Recall note 8 and think of the mood being generated by popular works like *The Uninhabitable Earth* or those of Rifkin mentioned in note 49. Even capitalism's invasion of the Green Movement can thus be seen as seeding its collapse, especially if nudged by a freshened ethos of E. F. Schumacher's slogan-title *Small Is Beautiful* of 1973 (London: Harper & Row). But there must be a push, a very personal pushing, by you and me, to reinvent Schumacher's subtitle, *A Study of Economics as if People Matter.* This is a massive issue that I have avoided here, but I deal with it elsewhere, most recently in "Finding an Effective Economist: A Central Theological Challenge," *Divyadaan: A Journal of Philosophy and Education*, vol. 30, no. 1 (2019), 97–128. In that volume, too, there is an identification, in four other articles of mine, of a potential converging alliance of world religions on the present cultural crises. The converging is to involve their escape from the subtle tentacles of capitalism—noted by Marx and Ruuska—into a freedom from stale self-preservation. Might I say that, thus positively and intellectually converging on Gaia, religious capitalism is to move, so sadly slow, to be on our side? The title of the 2019 *Divyadaan* volume I refer to is "Religious Faith Seeding the Positive Anthropocene Age."

[33] We return here to the problem expressed in note 3 and in a weave of the notes named there. Science leans us forward willy-nilly. What we need, flowing in our bones, is the large genetically-bent science of history that acorns tell us of, that I tell you of presently, as I lean here, a weak companion to an old oak.

But we should be tuned to the fact that we are, in some way, hovering over the middle zones, D_i and F, of my suggested steps, reaching for a redeeming of our attitude towards the full human story. To those zones I return in sketchy detail in Part Two. Here I poise us over the apparently simple dynamics of genetics.

In this poising I lift us, indeed, to both musing and being in a simple dynamics. To our shared question, "to what shall I compare / the world and human life?" I suggest: to a sunflower, a sunflower gently questioned: "Sunflowers Speak to us of Growing."[34] In what odd sense can this Gaiac appeal bring us into a simple isomorphic dynamics? Let us view that wondrous dynamics of a sunflower's week-by-week growth as paralleled by our era-by-era growth. I have seen a sunflower battle gallantly through drought and weeds in the early weeks to come out finally with its glorious yellow and brown grin. There is an evolutionary dynamic that weaves it forward. What is that dynamic? It is a developmental bent. But what, pray, is that? "Does not everyone have some notion of what development implies? Undoubtedly most of us have. But when it comes to formulating these notions, they turn out to be very vague."[35]

Shift now from the sunflower weeks to the earth's eras.[36] We gather as a group here, as other such groups gather in these decades, with a genetic bent that is "very vague." The sunflower's bent is spontaneous, as is ours. But ours leads us to gather questionings. The gatherings are seeded by a spontaneity that somehow is a questionable spontaneity: that indeed is both the content and the ethos of our gathering. So, capitalism is questionable, but it has a spontaneity that we question as foreign to fulsome flowering. What is that fulsome flowering? Answers from anywhere are "very vague." There is no parallel, in that perspective, to the brown and yellow grin.

We identify the weeds of capitalism in the garden of our era, but we too are in that garden, and with a spontaneity that parallels the sunflower's bent. Toni Ruuska, in that garden, identifies the weeds but also flickers of sunflower sanity.

---

34 The title of my *Cantower* 2, a website series 2002–2012 of 150 essays inspired by Ezra Pound's series of 117 *Cantos*. The series is available at: http://www.philipmcshane.org/cantowers.

35 Paul Weiss, *Principles of Development* (New York: Henry Holt and Company, 1939), the introductory sentence.

36 In reflecting over either flowers or florists or our futures you need to battle over an academic reductionism, e.g., a crazy mythology of genes. Such a view is represented by Part VI, "Emergence, Life, and Related Topics," *Science and Ultimate Reality: Quantum Theory, Cosmology and Complexity*, edited by J. D. Barrow, P. C. W. Davies and C. L. Harper (New York: Cambridge University Press, 2004).

Is the identification complete? He does not claim that it is. Still, haunting his incomplete identification is a "very vague" genetics **both of** the twisted human spontaneity that, indeed, can be "the general direction imposed on social life by the dominant fundamental group" of twisted people **and of** a deeper spontaneity that sniffs the rot.

I leave the problem of a struggle to identify that deeper spontaneity to Part Two, but here I wish us to reach for an identification of incomplete identification as that incompleteness cramps all our efforts.

What is desperately needed, be it in botany or in logic or in reaching for an effective view on sustainability, is a standard model of genetics that would parallel the standard model that stabilizes and directs present physics. And there, indeed, ay, is the rub.

The weeds cripple the "very vague" genetic climb: is that not the message of *Reproduction Revisited: Capitalism, Higher Education and Ecological Crisis*? And can we not detect, very vaguely, that the detection of both the weeds' damage and the seeds' survival is very vague? Can we, further, dream of, imagine into, its replacement as an effective step forward? So that we replace the vague struggle against the negative Anthropocene as the positive heart of our endeavor? So that we could come to speak a fresh ending to Ruuska's book, of the controlling genesis of genetics as what we need to sniff out how we have soiled the riverruns of the world, to sniff out how to turn to their cleansing adequately. Thus there emerges a luminous foundational community that takes the stand, "It *can* be replaced. It *must* be replaced. It *will* be replaced"? We can meantime face, in skirmishing fashion, the dress and address of doom that closes in swiftly on our lungs and hearts and trees and rivers. But the novel genesis of a novel genetic control of progress needs to be at the heart of our reach forward, short term and long term. A shabby standard model of genetic progress, and an increasingly enlightened ethos of its shabbiness, must be at the heart of a cyclic collaboration, a genetic identification of each specialized group-step of the search for sustainability and peaceful coexistence.

Part Two: Remembrance of Times Past and Future

My thesis is simple. Effective progress in the global move forward requires that that movement face the challenge of shifting, in this century, from developmental vagueness to the precision of a genetic heuristics, through the gradual emergence of the functional division of labor skimpily diagrammed in the step diagram. But my thesis is "very vague" to you, and the slight clarity that I have achieved in the past fifty years is not something to be communicated in a short chapter. Ay, there's another rub! This chapter is, as I claimed, an essay in the functional zone of C_{59} diagrammed in my third diagram above. It is, of necessity, a popular communication. Fortunately I was nudged towards a decent strategy of communication by two books familiar to the present group: Toni Ruuska's book, and the book *Sustainability and Peaceful Coexistence for the Anthropocene*. Paolo Davide Farah's Foreword to the latter book leads me forward. "The crucial role of human beings and their activities in the multiplicity of crises in the current world – ranging from ecological to economic and socio-cultural – cannot be disputed, but their complex character requires the adoption of a holistic approach to the problematic issues."[37] The holistic approach—one that does not emerge in that book but the desire for it lurks there—is a future thing, a dream-goal that I would obviously like shared, even as we venture forward in these decades in semi-effectual skirmishes.

Ruuska's book gives me leads forward in my communication. Even a casual pause over his first chapter hints at a distant wholesome genetics as it hovers over Marx's ontic genetics of the shadow of such a phyletic genetics. I am, however, aiming here, not at a summary or a critique, but at us sniffing out features of his effort that can help us forward to the sniff of a vision: might I say a 2020 vision?

The issue is a sifting out of the past a genetics of progress, something that Ruuska edges towards right along in his chapters. He picks his way through the particular threads of the historical process clearly named in his title, but it is worth our pedagogical while to pause in his chapter on Marx. There he weaves his own search for a stance round the search of Marx and for Marx's sequence of stances. In both Ruuska and Marx the personal element is central:[38] the Hamlet element, as I call it, an elemental questioning in search of a life for Pericles and posterity. Gradually there emerges the integrity of the search, climbing through the ten pages 35–45. The search is for a "science of history,"[39] a "dialectical whole,"[40] a unity of

---

[37] *Sustainability*, xiii.

[38] Obviously, the mood is set by Ruuska's Introduction, and it persists right through to his final vehement stand on the death of capitalism. In between there is a genetic persuasive weave at the center of which is his nuancing of Marx's stand. Part of that persuasion is the genetics of his climb. This chapter puts such personal poising in the fuller context of what Eric Voegelin calls, in obscure suggestiveness, "the dialogue of humanity with its humility" (see note 68). But it asks you, as does Ruuska's book, to take a stand.

[39] *Ruuska*, 35.

[40] *Ruuska*, 37.

the human and the ecological,[41] a unity that detects a core flaw in humanity's western education, it being "an instrument to spread bourgeoisie moral principles."[42]

Here I pause, returning to an earlier pointing regarding Aristotle: in the words of one of my mentors in all this, "Aristotle was a bourgeois."[43] To talk of a science of history in Marx's sense is to step clearly away from Aristotle: indeed it is a step beyond Hegel towards a poise on the meaning of *is*.[44]

Before I venture into this odd claim and Ruuska's neat positioning on ontology, I wish to lead us to a very elementary observation that is nonetheless of basic significance for my cultural lead to a novel division of labor. The chapter on Marx is clearly a chapter on Interpretation, and one could even see it as primarily a discourse of the type C_{22} in the display of the third diagram above. To analyse Ruuska's strategies of weaving back and forth on the emergence of Marx's writings and the resulting spectrum of interpretations would be quite a lengthy undertaking. But is it not fair for me to note Ruuska's dependence on a first community, C_{11}, of researchers? The division of labor I write of is thus present, as it is, so evidently, in modern physics' split between the investigators of particle tracks and the overlay of those who interpret the stuff into theory.

Further, I wish to add the suggestion borne out as we move along through the book, that Ruuska's interest is in "telling the story like it is,"[45] all the way to his

---

[41] Marx emerges, as "an important ecological thinker" (*Ruuska*, 39). See ibid, 101–104.

[42] *Ruuska*, 41.

[43] The mentor is Bernard Lonergan (1904–1984), who presented me with the structure I write of here in an afternoon conversation of the summer of 1966. The quotation is from a ten-page letter he wrote to a Jesuit Superior in January 1935. The letter is reproduced Pierrot Lambert and Philip McShane, *Bernard Lonergan. His Life and Leading Ideas* (Vancouver: Axial Publishing, 2010), 144–54. The quotation is on page 152.

[44] Ruuska raises issues of ontology and epistemology on pages 30–31. I pass over them here, because it seems to me that they are beyond the communal reach of axial humanity, so luminosity regarding and guarding is—is?, is!, is.—is in the zone, in this millennium, of evolutionary sports, a luminosity certainly beyond the reach of axial men like Aristotle and Hegel.

[45] This is a question that pushes us, as Ruuska pushes us, to view the story with his leaning towards his final cry against capitalism. I add a fuller context in the Epilogue. There is a slender deliberative poise in his telling. It is symbolized in its fullness by my image of a leaning tower: but what a feeble printed image. To be surrounded with a molecular symphony: that is our task this millennium. But you might muse, vortex-wise, over, e.g., the trail from note 3, finding the nudges towards a deliberative poise for this

identification of Finland's dynamics of higher education. He moves, thus, through a zone of recent history[46] to come up with his view. Is he thus not into the ballpark of **H**, of History? Is there an obvious yes answer to this? There is, yet it needs qualifications, qualifications indeed that can throw wondrous light on our entire project. But first I wish to sweep past that problem of qualifying, in a creative positive way, Ruuska's subtle achievement and deal more simply with what he and I and we are at, are reaching for.

So, sweeping through his chapters here I suggest that he moves forward to venture, in a loose fashion, beyond **H** to **D$_i$**: his poise is critical, dialectical, and we shall muse further about it. But for the moment I wish to hurry us on to the ending of the book, a glorious moment in my first reading of it. He ended, to my amazement, where my own theoretics of Dialectic, of **D$_i$**, ends. For me it is a clear baton-exchange in the relay that is the ten-step run, the baton-exchange to a community facing the future foundationally. "Capitalism is a historical structure. It *can* be replaced. It *must* be replaced. It *will* be replaced."[47]

But I had best quote his admirable ending more fully. I am climbing towards putting it in a startlingly new context, to end part two in an effective strangeness.

> From an ecological perspective, anti-capitalism in education, or in any other level of social organising, is not to be considered radical, but in fact plain common sense. Sadly, it is clear that the current mental mind-set deems anti-capitalism revolutionary. In contrast, capitalism is very destructive ecologically, but socially legitimate, at least for the time being. This is why I have attempted to portray capitalism the way it is:

---

morning—"I caught this morning morning's minion, kingdom of daylight"—and this millennium. I add there, here, the beginning of the flight of G. M. Hopkins' *The Windhover* to my plea. We are at a ridge, a sillion, in history. Do his final lines not invite, my chevalier, my dear: "And the fire that breaks out from thee then, a billion / Times told lovelier, more dangerous, O my chevalier / No wonder of it: sheer plod makes plough down sillion / Shine and blue-bleak embers, ah my dear, / Fall, gall themselves, and gash gold-vermilion."

[46] What is recent history? One may think of the Industrial Revolution, the emergence of modern pseudo-economics and crippled education, and so bring a context to and for Ruuska's work. But one should be prepared to tune into his leaning forward in the venture, the leaning forward talked of in the previous note. And then there is the context offered in my Epilogue, that, strangely, pushes us to a serious molecular hopefulness, that the recent is just the weediness of our season in axial hell: our sunflower is to turn out of the tangle and smile.

[47] *Ruuska*, 254.

a radical utopia running against the foundations of life. Especially from this perspective, Karl Marx is truly an important thinker and historical figure. He famously pointed out that societal structures and institutions are not eternal, and argued instead that any historical structure can be transformed or replaced (Eagleton, 1999). Capitalism is a historical structure. It *can* be replaced. It *must* be replaced. It *will* be replaced.[48]

The notion of relay has been with me for decades and it is a clearly encouraging nudge: ten running against one in a 10,000-meter race is no contest. But perhaps more significant for giving a notion of future functional collaboration is noting what I call the shift here from pin to pen, or I began calling it yesterday, the shift from pin to PEM.[49] Was the shift from domestic pin-making to the collaborative dynamics of the pin-factory, the automobile-making cavern, really genuine progress? Not our question for the moment. Our quest is leading us to see that, yes, pen-using in the cyclic step dynamics I have sketched is the seed of Progress Effectively Mantelled. Is to be or not to be? I steal a line just prior to Hamlet's entry, to you now entering my writing, my speaking: "How smart a lash that speech doth give my conscience."[50] Do I give you pause here? Might I give you poise here? I think of my crazy fellow Irishman meeting and greeting Plotinus as he then turned to the task of translation, writing in his diary at age 38, "This is worth a life."[51] Might we not translate Ruuska's play of words into a whirl of effective anti-capitalist education?

I will write more about that translation and its effective contextualizing in the Epilogue. But now I wish to share the high point of my adventure with his book: an identification of his play, his audience.

In his first personal chapter he writes, "I attempt to convince the reader."[52] Let us pause and puzzle about his reader, his convincing, his attempt. In the

---

[48] *Ruuska*, 254. The internal citation is to Terry Eagleton, *Marx. Great Philosophers Series*, (London: Routledge, 1999).

[49] PEM: "**P**rogress **E**ffectively **M**antelled": recall note 16 above. My neologistic move to verbalize *mantel* points to a quite different world than that of Drucker's management: or—I happen to have at hand the semi-pop truncated writings of Jeremy Rifkin— lightweight shiftings to *The Emphatic Civilization* (New York: Jeremy P. Tarcher, 2009) or *The Third Industrial Revolution* (New York: Palgrave Macmillan, 2011). Now call in, haul in, note 32, with something of the crazy mood of note 45 above.

[50] Shakespeare, *Hamlet* III. i. 53.

[51] Stephen McKenna produced a magnificent translation of *The Enneads*.

[52] *Ruuska*, 13.

present conventions of writing and reading, his readers are an indeterminate audience, his convincing is a matter of, well, us or "them," his attempt brilliant for us but, for "them," an illusion about their solemn comedy of errors.[53] For us and from us? There is, I would assume, applause, and an agreement continuous with the agreement we have about this and the two prior *Peaceful Existence Colloquiums*. On the other hand there may be a fresh identification,[54] a fresh hearing, a fresh view of the mark that Marx made in viewing capitalism, a fresh view of Ruuska's view, startling for him, for me: and how, perhaps for you, perchance a dream, ay, a psychic rub?

What, then, THEN,[55] if his attempt was identified as a great shot at the functional step of Interpretation? Such a functional interpretation lives within a history of interpretations and aims—I think of a maturity of the project—at contributing to a lift of that history, an effective lift. The mature contribution passes on, baton-wise in two senses of wise, from the history of ideas to the history of flawed achievements. And so on, where that "so on" means a "sow on" creatively spreading through the sequence of collaborative communities pointed to by D_i, F, D_{oc}, S, C.[56] What more can I say?: there are volumes to be thought

---

[53] Them? : "wanting guilders to redeem their lives" (Shakespeare, *A Comedy of Errors*, I. i. 8).

[54] More details on this fresh identification are in my *Cantower* 3, "Round One Willing Gathering," section 3, "Identifications." (http://www.philipmcshane.org/cantowers) The transition problem we have is to tune psychically in the step-working symbolized in my second diagram. Ruuska motivates that tuning. Without the tuning we march towards *The Uninhabitable Earth*.

[55] I recall Marcuse on fantasy (see above, note 18). My *Cantower* 5, "Metaphysics THEN" begins with a last poem by Samuel Beckett: "go where never before / no sooner there than there always", and weaves into the spread of a Scottish love-song both Ezra Pound and George Sand. Both these strange people invite us to the psychic attunement mentioned in the previous note. "upon the gilded tower in Ecbatan / Lay the god's bride, lay ever, waiting the golden rain" (Pound, *Canto* IV). "The consciousness of self as animal, vegetable and mineral, and the delight we feel in plunging down into that consciousness, is by no means degrading. It is good to know the fundamental life at our roots, while we reach out towards the higher life which is completely attained only in flashes of insight and in dreams" (George Sand, 1952, *The Intimate Journal of George Sand*, translated by Marie Jenney Howe, Haskell House, New York, 1975, 182). "The Great Shot's" roots are here.

[56] I end here the set of footnote pointers that stretched through notes 3, 6, 17, 20, 22, 33, 45. But, obviously, I make a somewhat arbitrary pick of the chords to the melody of my text. What, then, is deliberation? It is an unknown of history, an *X* I may call

out and implemented in these next centuries of anti-capitalism.[57] At this stage in my writing I returned to Næss's work for the first time since I read it thirty years ago in Oxford. It still astonishes me. "For Næss, Deep Ecology is not a rigid dogma, but rather a 'platform' that draws together supporters from disparate backgrounds and gives them a base from which to reassess humanity's relationship with Nature."[58]

So I am led to halt this Part Two abruptly, poising you before a new version of Næss's invitation. Ruuska freshens our grip on higher education, a busy idleness that is a sell-out to the sickness he identifies, "the unyok'd humour of your idleness," as I now call it, in the recollection of a soliloquy that leaped out at me, yes, seventy years ago. It is not a matter of walking away from a sick *Paideiad* but of Trojan horsing it towards a quite new being. Might you take a stand with me?

> I know you all, and will awhile uphold
> The unyok'd humour of your idleness;
> Yet herein will I imitate the sun,
> Who doth permit the base contagious clouds
> To smother up his beauty from the world,
> That when he please again to be himself,
> Being wanted, he may be more wondered at
> By breaking through the foul and ugly mists
> Of vapours that did seem to strangle him.[59]

Epilogue: The Strangled Beauty

The strangled beauty is evolution's sown what.

Cosmopolis. I think now of a previous puttering of mine in Puebla, Mexico, 2011. In my talk, "Arriving in Cosmopolis" (http://www.philipmcshane.org/website-articles), I put that arrival at 9011 A.D. I even spelled out the population percentages in each of the groups $C_{i,i}$. Will seven millennia prove me right? See, there it is, the mark: the question mark, the mark in your molecules which is your sharing of the Compelling Genius of History!

[57] In my notes (see, e.g., the recent note 55) I have been nudging us to a fantasy of a strange future, but its beginning is a beginning of a new integral thinking that is, in a deep sense, a recovery of primitive integrity, the achievement of the cosmicauled lonely molecules of the *moi intime*.

[58] I quote the pre-comment on his article, "Deep Ecology and Ultimate Premises," *The Ecologist*, vol. 18, no. 4/5, 1988, 128, available at: https://www.resurgence.org/magazine/ecologist/issues1980-1989.html.

[59] Shakespeare, *Henry IV, Part One*, I. ii. 188–96.

The emergence of humanity is the evolutionary achievement of sowing what among the cosmic molecules. The sown what infests the clustered molecular patterns behind and above your eyes, between your ears, lifting areas—named by humans like Broca and Wernicke—towards patterned noise-making that in English is marked by "so what?"[60]

Eric Voegelin, in the concluding chapter of the second last volume of his *Order and History*, raises in its fullness Marx's question of a "science of history."[61] "The 'absolute epoch,' understood as the events in which reality becomes luminous to itself as a process of transfiguration, is indeed the central issue in the philosophy of history."[62] It is the issue that is raised by me implicitly in the quotation with which I begin this Epilogue, identifying, if you like, Ruuska's "provoked serious questions"[63] as the heart of the problem in its full hearty sense. What is not luminous to itself, so its group-evolution begins with a stumbling: what is what is not an issue; the issue becomes, slowly, what might gather berries. "What defines a man?"[64] inquires Arjuna, and Chrisna does not answer, "Yes, what is man."[65] Nor do we, in any serious luminosity to itself, our selves, what vague, not bright in what's ayes.

There was, then, Karl Jaspers' faulty answer of identifying B.C. 800–200 as an axial period of history: luminous differentiations occurred in Greece, Persia,

---

[60] Philip McShane, *The Allure of the Compelling Genius of History: Teaching Young Humans Humanity and Hope* (Vancouver: Axial Publishing, 2015), 3: the beginning of chapter 1, "Sow What." Paul Broca and Carl Wernicke were 19th century physicians who discovered that impairments in understanding speech and speaking were linked to damage to specific parts of the brain and set off further work on the localization of brain function.

[61] *Ruuska*, 35.

[62] *The Ecumenic Age*, 381.

[63] *Ruuska*, 46.

[64] I am quoting from the translation by Barbara Stoler Miller, *The Bhagavad-Gita: Krishna's Counsel in Time of War*, (New York: Bantam Books, 1986), The Second Teaching: Philosophy and Spiritual Discipline, Verse 54, page 39. I go into detail around this point in section 1.4 "*Bhagavad-Gita*: Song of the Adorable," in *Process: Introducing Themselves to Young (Christian) Minders* (available at: http://www.philipmcshane.org/website-books).

[65] The reply, however, is complex and richly suggestive. For example, from Book II: "When he gives up desires in his mind / is content with the self within himself, / then he is said to be a man / whose insight is sure, Arjuna" (Verse 55) and there is the seed of the rejection of initial meanings: "Undiscerning men who delight / in the tenets of ritual lore / utter florid speech, proclaiming, / 'There is nothing else!'" (Verse 42)

Israel, India and China.[66] Arnold Toynbee took issue with Jaspers and suggested a larger spread of centuries.[67] Voegelin noted the parallel between the poise of the Sumerian King List and Hegel's philosophy of history and thus shook up the spread considerably.[68] My own strange view, already mentioned, spreads the struggle to beyond our times by identifying the muddled climb of evolution's aggregates of whats as trailing forward towards what I call a positive Anthropocene,[69] when, yes—ay, there's the rub!—humanity becomes (beyond Poisson to Bell-curve)[70] effectively self-luminous. Within that period what blossoms into what is called *religiosity*, and that religiosity too stumbles along with

---

[66] Karl Jaspers, *The Origin and Goal of History*, (London: Routledge & Kegan Paul, 1953), chapter 1.

[67] Arnold Toynbee, *Mankind and Mother Earth: A Narrative History of the World*, (London: Oxford University Press, 1976), 178: he insists on enlarging the axial period to seventeen centuries, including thus Zarathustra and 'Deutero-Isaiah,' Jesus, and Muhammad.

[68] Eric Voegelin, *The Ecumenic Age*. "And what is modern about the modern mind, one may ask, if Hegel, Comte, or Marx, in order to create an image of history that will support their ideological imperialism, still use the same techniques for distorting the reality of history as their Sumerian predecessors?" (118). "A 'modern age' in which the thinkers who ought to be philosophers prefer the role of imperial entrepreneurs will have to go through many convulsions before it has got rid of itself, together with the arrogance of its revolts, and found the way back to the dialogue of mankind with its humility" (252).

[69] For an imaging of the convulsions of which Voegelin writes, see my essay in chapter V, "Ant Hopper." What of the dates-problem here and in the general discussion of the Holocene and its overlapping with the Anthropocene? The problems raise issues of classifications that are best skipped here. A sketchy indication of my view puts the Anthropocene further back even than the emergence of language, thus an early beginning of what I call the negative Anthropocene: a stumbling spontaneity. The turn that invents grammar—I think of Sanskrit's efforts—also invents the grounds of distortion away from the seed of history. Think of the place of interrogatives in your own present grammar. So, Axial Aristotle weaves forward in a possibility of a stage of the negative Anthropocene: subjectivity is truncated, blocked by its own objective eloquence. In this state we live and move and have our being: blocked heads of state, blocked heads of corporations, blocked heads of university departments, blocked heads on television.

[70] I have in mind the shift of Fisher's heuristics of history referred to earlier: see note 16 above. But the statistics of recurrence-schemes in glocal situations is altogether too complex to even hint at here.

faint glimmers of luminosity.[71] *Upanisad* "has been explained etymologically as the teaching obtained from sitting (*sat*) devotedly (*ni*) near (*upa*) a teacher"[72]: the evolutionary what has yet to sit near Gaia or the lightsome what in Gaia's crown of neurochemicals. Whitson wrote of *The Coming Convergence of World Religions*,[73] but the convergence he wrote of is a truncated thing, continuing to sit apart from Gaia. My recent effort points to a turning that is a boost to the fuller human search for what I call a new Han Dynasty, "Step-Han finds his Mother."[74] The passive *convergence* becomes the active *converging*, thus sharing our spontaneous whatting bent that spontaneously shares the long-eyes view of Voegelin,[75] and to be called upon, called out, called in, called into the *moi intime*, into the fullness of Gaia's dynamics. "Each member, each group, indeed our whole host and its great pilgrimage, was only a wave in the eternal stream of human beings, of the eternal strivings of the human spirit towards the East, towards Home."[76]

---

[71] Recall my comment on luminosity in note 44 above. The self-luminosity at the— and in the—heart of it is the stumbling goal of the Axial Period, which separates the two times of humanity: the time of unknowing spontaneity of whatting with its present ant-heap stage, and the time to come of self-luminous control. We reach, grasshoppers, for that time, now, ontically and phyletically.

[72] Richard V. De Smet, *Guidelines in Indian Philosophy*, chapter 3, "The Upaniṣadic Discoveries. 1. The Quest for the Brahman," *Divyadaan: Journal of Philosophy and Education*, vol. 21, no. 2 (2010), 257.

[73] Robley Edward Whitson, *The Coming Convergence of World Religions* (New York: Newman Press, 1971).

[74] There are several relevant implicit references here. First, I refer to *The Allure of the Compelling Genius of History*, where I write of a new Han Dynasty quite beyond that old Axial Han Dynasty of 206 BC – AD 220. (See the back cover and also 202, 230–2). But there is a heuristically-telling reference to Margot Norris, "The Last Chapter of *Finnegans Wake*: Stephen Finds His Mother," *James Joyce Quarterly* vol. 25, no. 1, 1987, 11-30. On page 11 Norris writes, "Using the device of *anastomosis*, Joyce attempts, in the last chapter of his last work, to bridge all the great ontological chasms." Think out *Ana-*, again, *stomein*, to provide with the mouth, in terms of integral global subjectivity speaking luminously to itself in a distant dance of humanity. And perhaps read yourself into *The Well of Loneliness*, Radclyffe Hall's 1928 novel introduction to you of the lonely lesbian Stephen.

[75] I originally treated this problem in chapter 1, "Middle Kingdom: Middle Man. T'ien-hsia: i jen" of *Searching for Cultural Foundations*, edited by Philip McShane (Lanham: University Press of America, 1984).

[76] Herman Hesse, *The Journey to the East* (New York: The Noonday Press, 1970), 12. I conclude here with a quotation with which I began the Epilogue, "Being and Loneliness," of *Wealth of Self and Wealth of Nations: Self-Axis of the Great Ascent*, a book

which was a revision of the book, *Towards Self-Meaning*, written with my colleague Garrett Barden (Dublin: Gill-McMillan, 1969). That was what I would now call my first *Common Quest Manifesto*. My final, vigorously critical "A Common Quest Manifesto" appeared recently, as *Æcornomics* 3, http://www.philipmcshane.org/ecornomics.

Afterword

James Duffy

There's far too many of you dying
You know we've got to find a way
To bring some lovin' here today, yeah

. . . .

Oh, what's going on (What's going on)[1]

What's Been Going On?

In May of 2019, a month before the *The 3ʳᵈ Peaceful Existence Colloquium* took place in Helsinki, Finland, the 34-member Anthropocene Working Group (AWG) voted to use the term *Anthropocene* to designate a new geologic epoch, with a majority twenty-nine members voting in favor of beginning the new geological epoch in the mid-twentieth century. Currently there are 12 teams of geologists doing field work on five continents looking for a 'golden spike'—a distinct and unmistakable marker that will pinpoint the birthplace of the Anthropocene, a new epoch of geological time. The sites that are being considered include a cave in northern Italy, corals in the Great Barrier Reef, a lake in Ontario (Canada), another lake in China, and the Florida Keys on the Eastern Gulf of Mexico. It is projected that by the summer of 2022, a golden spike will be selected from the various sites currently being investigated. After that it will likely be another two or three years before the term is formally approved.[2] The hope is that this ongoing procedure to formalize the meaning of *Anthropocene* "can positively contribute to this cross-disciplinary debate and help achieve clarity in the use of the term 'Anthropocene.'"[3]

---

[1] Marvin Gaye, *What's Going On?* (Motown Records, 1971). This song was released in May of 1971 by Motown Records. In 1990, after being freed, Nelson Mandela recited its lyrics at Tiger Stadium.

[2] Initially it is for the Anthropocene Working Group (AWG) to decide what should be the primary marker or the main sign of human fingerprint worldwide. After a site is chosen, the Subcommission on Quaternary Stratigraphy (SQS), a subcommittee of the International Commission on Stratigraphy (ICS) will vote on alternatives, and then the ICS itself would have to approve. The final step is ratification by the International Union of Geological Sciences.

[3] Jan Zalasiewicz et al., "The Anthropocene: Comparing Its Meaning in Geology (Chronostratigraphy) with Conceptual Approaches Arising in Other Disciplines," *Earth's*

What else has been going on in the last three years? As I type, an estimated 6.25 million people have died from COVID-19, which has impacted daily rhythms of work, study, and leisure around the globe. Governments and corporations mobilized their resources to create vaccines while billions of people discerned how to best deal with the virus. Amid a lockdown in March 2020, quarantined Italians sang from their balconies, and the video went viral.[4] In 2020 China began asserting itself and dashed hopes that it would become "a responsible stakeholder" in world politics. In the same year, European Union renewables overtook fossil fuel-powered energy for the first time, a milestone in Europe's Clean Energy Transition.[5] On July 1, 2020, Philip McShane, who was nominated twice for the Templeton Prize for his steadfast work in foundations of mathematics, natural science, economics, philosophy, theology, and omnidisciplinary methodology, and who coined the phrase "Openers of the positive Anthropocene,"[6] passed on to an everlasting neurodynamic state of "infinite surprise."[7] On January 28, 2021, Paul J. Crutzen, who was awarded the Nobel Prize in chemistry in 1995 for his work on the Earth's atmosphere and was once the world's most-cited scientist, passed on.[8] In February the US officially rejoined the Paris Agreement. and a report revealed that, for the first time, renewables generated more electricity than fossil fuels in Europe.[9] In March of 2021, in the midst of the pandemic, Pope Francis met with Grand Ayatollah Ali al-Sistani in Iraq, which was the first ever

---

Future 9, no. 3 (2021), e2020EF001896, https://doi.org/10.1029/2020EF001896. Hereafter "The Anthropocene: Comparing Its Meaning."

[4] Nearly three million people have watched the video, "Coronavirus: quarantined Italians sing from balconies to lift spirits,"
https://www.youtube.com/watch?v=Q734VN0N7hw.

[5] Dave Jones, "EU Power Sector 2020, Electricity Trends," Ember, January 24, 2021, https://ember-climate.org/insights/research/eu-power-sector-2020.

[6] See "The Masses and Sustainability," 111.

[7] The last two words in the epilogue to Philip McShane, *Wealth of Self and Wealth of Nations: Self-Axis of the Great Ascent*, 2nd ed., ed. James Duffy (Vancouver: Axial Publishing, 2021), 95. See also "Prologue: The Betweenness of Death" in Philip McShane, *The Everlasting Joy of Being Human* (Vancouver: Axial Publishing, 2013), 1–12.

[8] *Paul J. Crutzen and the Anthropocene: A New Epoch in Earth's History*, Susanne Benner, Gregor Lax, Paul J. Crutzen, Ulrich Pöschl, Jos Lelieveld, Hans Günter Brauch (eds.) (Switzerland: Springer, 2021). A list of his contributions to this book is available at https://link.springer.com/book/10.1007/978-3-030-82202-6#toc

[9] "Digest of UK Energy Statistics (DUKES)," Government Digital Service (GDS), https://www.gov.uk/government/collections/digest-of-uk-energy-statistics-dukes

meeting between a pope and a grand ayatollah. In August of 2021, UN Secretary General António Guterres described the UN report as "a code red for humanity," underscoring the catastrophic change in climate if emission of heat-trapping gases is not reduced. *The 4th Peaceful Coexistence Conference* took place virtually the first week of December 2021. A little over two months later, Russian President Vladimir Putin launched an invasion of Ukraine, sparking the biggest war on a European state since World War II, one that could drag on for many months. On May 10, 2022, Andy Warhol's painting of Marilyn Monroe sold for $170m ($195m including taxes and fees).

The list of events is somewhat random and at best impressionistic. Why is the sale of Andy Warhol's painting of Marilyn Monroe there? Why are migration crises and shortages and stoppages of supply chains not on the list? Was the election of President Biden in November of 2020 significant? Adding to or subtracting from the list of events would merely create another list, more names and dates. In the last three years, academics and other professionals have gathered, exchanged papers and ideas, and co-authored articles and books. Have the tens of thousands of publications on the Anthropocene published between March 2019 and March 2022[10] led to an increment of progress? Or does "big data just keep getting bigger"?[11]

A Conundrum and a Proposal

The collection of essays in this book emerged over three years ago when the four of us read and reacted to *Sustainability and Peaceful Coexistence for the Anthropocene*,[12] co-authored by 12 individuals. Since then, other individuals have come together to co-author books on the Anthropocene, for example the three authors of *The Anthropocene: A Multidisciplinary Approach*, who are among the 27 co-authors of

---

[10] In April of 2019, a search of the Library of Congress catalog for the keyword "anthropocene" returned 480 items, mostly books. A search in May of 2022 returned over 1,300 items. Members of the International Panel on Climate Change (IPCC) estimated that the relevant new literature on climate change between 2016 and 2018 "was somewhere between 270,000 and 330,00 publications" (Julia Adeney Thomas, Mark Williams, and Jan Zalasiewicz, *The Anthropocene: A Multidisciplinary Approach* [Cambridge: Polity Press, 2020], 7). "Grappling with even one factor of the many that make up the Anthropocene requires the labor of thousands of scientists and extremely powerful computers." Ibid, 6.

[11] *The Anthropocene: A Multidisciplinary Approach*, 7.

[12] See note 10 in the Preface.

"The Anthropocene: Comparing Its Meaning in Geology (Chronostratigraphy) with Conceptual Approaches Arising in Other Disciplines."[13] Both the book and the article advocate collaboration and indeed invite it.[14] This, I believe, is a crucial first step. A second, more difficult step is to figure out how to divide up a multitude of tasks in order to collaborate efficiently. As I wrote in the Preface, "the lack of coordination is one of the pain points of our efforts, all our efforts."[15]

The three authors of *The Anthropocene: A Multidisciplinary Approach* express the pain point in terms of a meaning conundrum. Since people in different disciplines work on different time scales and embrace different methods of research (including different citations styles), multidisciplinary conversation is tenuous. "We are not always talking about the same thing, or talking about it in the same way."[16] While some might consider meaning a secondary matter, it is a practical matter for those endeavoring to collaborate. The authors of "The Anthropocene: Comparing Its Meanings," write this about managing of meanings:

> While some social scientists and humanists align their understanding of the Anthropocene with chronostratigraphic and ESS definition of this phenomenon (e.g., Angus, 2016), others choose to redefine it or invent alternative terms such as Thanatocene, Thermocene, and Capitalocene (see Hallé & Milon, 2020) to offer different models of explanation for the current ecological crisis, though some may include elements of distrust of science (in turn partially manufactured by political and corporate interests to give impetus to those who wish to reject scientific findings: Oreskes, 2019). It is not clear whether the formalization of the chronostratigraphic Anthropocene, should it occur, will have any impact on humanists, social scientists, and others who are not ready to engage with scientific approaches such as in chronostratigraphy and

---

[13] Mark Williams and Jan Zalasiewicz are also members of the Anthropocene Working Group.

[14] "We encourage further discussion of this particular issue [multidisciplinary study of the Anthropocene(s)], of name and identity, among the scholarly communities involved, so that precise communication and effective collaboration in this important and wide-ranging area [figure 3. The Integrative Anthropocene Concept *sensu lato*'] might be facilitated." Zalasiewicz et al., "The Anthropocene: Comparing Its Meaning," 19. "In the face of unprecedented challenges, we need the rigor of established disciplines to ensure expertise and to assess evidence, but we also need these disciplines to be self-reflective and to engage with work not just in adjacent fields, but in distant ones." Thomas, Williams, and Zalasiewicz, *A Multidisciplinary Approach*, 5.

[15] See page v.

[16] *A Multidisciplinary Approach*, 13.

ESS. It is therefore important to consider how these various meanings might be managed in practical terms.[17]

Throughout this article, the authors contrast the loose use of the term with "the tightly defined, geological, chronostratigraphic concept,"[18] normative approaches with descriptive approaches, and fields that address issues of meaning and value with those that do not. What's needed, it is argued, is a balancing of "rigorous science" with the findings of other "complementary" cultural fields. So, on the one hand, the work of those intent on formalizing the term *Anthropocene* as a geological unit will be inadequate if "structured only around scientific findings and technological options."[19] On the other hand, "any response to the Anthropocene that is purely cultural without being rooted in scientific understanding" will also be inadequate. "Such emergent [non-geological] terms could comfortably sit alongside, and fruitfully interconnect with, the Anthropocene as proposed by Crutzen and now being explored by the AWG."[20] Dealing with the predicament we are in is "both a scientific and humanistic enterprise" even though "our joint efforts are very unlikely to produce 'grand integration' of all disciplines and worldviews."[21] The proposal includes an invitation to those who are not professional geologists or Earth System scientists to accept a formal definition.

The End of Innocence

How might the various individuals and groups concerned about the future of the Earth take responsibility for an indeterminate future? Not all are of the view that learning how to write and organize into religions based upon creeds was a "sweet spot for humanity."[22] Gandhi once replied to the journalist who asked him what he thought about Western civilization: "I think it would be a good idea." He had

---

[17] Zalasiewicz et al., "The Anthropocene: Comparing Its Meaning," 18. In this same article, the authors provide a table with examples of the use of the term *Anthropocene* from ten disciplines—geology, Earth System Science (ESS), geography, social science, archeology, anthropology, pedology, ecology and conservation biology, philosophy, history, and international law. The authors from diverse fields, including geography, geology, geological sciences, Earth sciences, history, chemistry, renewable resources, Arctic and Alpine research, and biology among others. "The Anthropocene: Comparing Its Meaning," 11–12.

[18] Zalasiewicz et al., "The Anthropocene: Comparing Its Meaning," 19

[19] "The Anthropocene: Comparing Its Meaning," 17

[20] "The Anthropocene: Comparing Its Meaning," 12.

[21] Thomas, Williams, and Zalasiewicz, *A Multidisciplinary Approach*, 17.

[22] Thomas, Williams, and Zalasiewicz, *A Multidisciplinary Approach*, 19.

a suspicion about measuring, or sizing up, what began to emerge in ancient Greece and Rome. The practitioners of Earth System Science, which is itself in its infancy,[23] adopt timescales as extensive as the processes and sub-systems they study, but they do not attempt to interpret or measure Western civilization. Historians of political science and philosophy typically do not think in terms of epochs, but some of them consider millennia or centuries, for example, the discovery of the noosphere in Greek literature and philosophy,[24] the shattering of Enlightenment hopes that the arts and sciences would promote progress in modern everyday life,[25] and the possibility that those of us involved in the modern academy suffer from a gross, undiagnosed arrogance.[26] "Does anybody really know what time it is? / Does anybody really care?"[27]

One telling sign of the times is an operative distinction between the two cultures that C.P. Snow described over sixty years ago.[28] "Confronting the Anthropocene is both a scientific and humanistic enterprise," so what is needed is mutual respect between those working in different fields of study, an acknowledgement of our differences, and a conscientious effort to balance scientific findings and those of the humanities and social sciences.

An image of the balance that comes to mind is that of an ellipse with two foci—rigorous scientific findings and those of complimentary cultural fields. If in our publications and working groups big and small, we travel around the ellipse

---

[23] Earth System Science emerged from a growing cultural awareness and concerns about environmental issues in the 1960s and 1970s and became a new science in the 1980s. Will Steffen et al., "The Emergence and Evolution of Earth System Science," *Nature Reviews Earth & Environment* 1, no. 1 (January 1, 2020), 54–63, https://doi.org/10.1038/s43017-019-0005-6.

[24] Bruno Snell, *The Discovery of the Mind in Greek Philosophy and Literature* (New York: Dover, 1948).

[25] Jürgen Habermas, "Modernity versus Postmodernity," *New German Critique* 22 (1981), 3–14.

[26] "A 'modern age' in which the thinkers who ought to be philosophers prefer the role of imperial entrepreneurs will have to go through many convulsions before it has got rid of itself, together with the arrogance of its revolt, and found the way back to the dialogue of mankind with its humility." Eric Voegelin, *The Ecumenic Age* (Baton Rouge: Louisiana State University Press, 1974), 192.

[27] Chicago, *Does Anybody Really Know What Time It Is?*, Chicago Transit Authority (Columbia, 1970). See also notes 4 in the Preface (p. ii) and note 11 in "Human Impacting Earth Systems," p. 14.

[28] C. P. Snow, *The Two Cultures* (Cambridge: Cambridge University Press, 1959).

and somehow maintain the proper tension,[29] communication is made possible. But if we do not, we will leave the orbit of the ellipse, thus losing the balance needed to sit alongside one another and dialogue. The thousands if not tens of thousands concerned about the future of the Earth face a practical problem of staying within the orbit of the ellipse, managing meanings "so that precise communication and effective collaboration might be facilitated."[30]

It is possible that "science for science's sake" and "art for art's sake" are enduring myths. That is another suspicion.[31] It is also possible that formalizing the term *Anthropocene* as an epoch, thus marking the end of the Holocene, might actually impede or delay rather than facilitate "a resolute and effective intervention"[32] in the Earth's processes, which—unless you conceive of *anthropos* as solitary, nasty, and brutish—includes processes like raising questions and offspring who are raising questions, as well as planning holidays, weddings, and cities. It is possible that the well-intentioned search for a golden spike in order to formalize the Global Boundary Stratotype and Section Point for the Anthropocene suffers from an eclipse of the mind and heart of *anthropos*.

Is defining the Anthropocene as a geological unit a step in the right direction? Those of us who do not understand what geologists and Earth System scientists are doing, as well as those who question why they are doing what they are doing, are being invited to accept a supposedly unambiguous meaning of *Anthropocene* now being explored by the AWG to stay within the orbit of the ellipse with its two foci—scientific understanding and the humanistic enterprise. This is part of a proposal for managing meanings.

I believe this proposal is somewhat naïve and that the image of an ellipse with two foci is misleading. The interpretation of rocks, time, and fossils, as well as the interpretation of the articles and books written about rocks, time, and

---

[29] For every ellipse there are two points, X and Y, called the foci, and a fixed positive constant d greater than the distance between X and Y, so that from any point P on the ellipse, the sum of the distances to the two foci (PX + PY) equals d.

[30] See note 14. Figure 3 divides disciplines into two groups, two levels: A consequential metalevel (social sciences, philosophy, cultural studies, psychology, history, anthropology, archeology, future studies, literature, gender studies, international law, etc.) and an analytical level (Earth System sciences, geology, paleontology, stratigraphy, geography, soil sciences.

[31] See further Philip McShane, "Aesthetic Loneliness and the Heart of Science," *Journal of Macrodynamic Analysis* 6 (2011), 51–84.

[32] Lonergan, CWL 18, 305–8.

fossils, is *de facto* conditioned by the horizon of the interpreter. There are a host of beliefs that contextualize the work being done, not just in the search for the golden spike, but in any field in any part of the globe. So, yes, "we need these disciplines to be self-reflective."[33]

"Science for science's sake" and its cousin "art for art's sake" are part of a familiar story, a cover story that permeates lower and higher education and was popularized in the television series "Big Bang Theory."[34] The end of innocence is the end of the innocent view that the research and interpretation of rocks or stars, fossils or fire, take place in a vacuum of 'pure science.'[35] Dating, describing, observing, analyzing, formalizing terms, and reaching conclusions are activities of an existential subject, someone who deliberates, evaluates, and decides. Primacy belongs to orthopraxis, not laws, codes, principles, or supposedly value-free formal definitions. There simply are no extrinsic criteria for determining who is an authentic or virtuous researcher or interpreter.[36] Moreover, my self-interpretation—which is implicit in interpreting any data—might reveal that "What is X?" at its best is "What might X be?" whether X is fire, running water, igneous, sedimentary, or metamorphic rocks, a sprouting sunflower, or a newborn child.

Those of us who are academicians have probably had candid students or colleagues ask at one time or another: "What does this seminar (essay, book, or paper for a highly specialized conference) have to do with 'real life'?" Well, whose real life? What do you mean when you make the noise *real?* The point is that my notion of "reality" or "real life" affects my performance, my doing, my conduct and, latently, my ethics of collaboration, indeed, my ethics of engineering a better

---

[33] See note 14. See also references to science—Aristotelian, primitive, pure—in the index.

[34] In the popular TV series Sheldon Cooper is a Caltech theoretical physicist who regularly chides his roommate Leonard for his work in experimental physics and his buddy Howard Wolowitz, who is an engineer at Caltech's Department of Applied Physics.

[35] In the Prologue to "Structuring the Reach Towards the Future," McShane makes the point by contrasting Archimedes' pragmatic scientific bent with Aristotle's (merely) speculative bent. See p. 117.

[36] Aristotle understood question-begging empiricism: virtuous people recognize virtue. "Virtue, then, is a habit, disposed toward action by deliberate choice, being at the mean relative to us, and defined by reason as a prudent man would define it." Hippocrates G. Apostle, *Aristotle's Nicomachean Ethics* (Grinnell, IA: The Peripatetic Press, 1975), 28–29.

life on Earth. The most fundamental question, the pain point of the negative Anthropocene, is about the future: Whether I am studying rocks or roses, ants or *anthropos*, where is my research, proposal, essay, or co-authored book "going" in the next thirty or three hundred years? To whom is it going? What might they do with it?

Rewilding Educating Rita to Get the Rhyme Wrong and Rita Right

In his book *A Life on Our Planet: My Witness Statement and A Vision for the Future*, the English biologist, natural historian, and broadcaster David Attenborough writes about a "hope that the wild can return even to land that was cultivated in Europe long ago."[37] In the conclusion to his book he writes: "I was born in another time. I don't mean this metaphorically, but literally. I arrived in this world during a period geologists call the Holocene, and I will leave it – as will every one of us alive today – in the *Anthropocene*, the time of humans."[38]

Attenborough's vision of recovery includes a re-wilding of the seas and land. His proposal for caring for the future of the Earth includes having fewer children, following a plant-based diet, flying less, and implementing a *circular economy* that over time removes any materials or chemicals from the economy. He writes of the importance of local communities taking responsibility for developing plans to increase biodiversity.

Not surprisingly, Attenborough does not write about rewilding playful, laughing, crying, longing humans, who are not outside of nature, unless you buy into such dualistic thinking.[39] Nor does he offer anything resembling the wild notion of *profit* as a "social dividend,"[40] much less a "global dividend," in his

---

[37] David Attenborough, *A Life on Our Planet: My Witness Statement and A Vision for the Future* (New York: Grand Central Publishing, 2020), 182. Hereafter *A Life on Our Planet*.

[38] A Life on Our Planet, 215.

[39] See the text cited at note 65 below.

[40] "The basic mistaken expectation rests on a failure to distinguish between normal profit which can be constant, and a social dividend which varies" (Bernard Lonergan, *Macroeconomic Dynamics*, CWL 15, 81). The social dividend is "income over and above 'standard of living,' 'rent,' interest, maintenance and replacement of capital equipment"; it is a "means given to entrepreneurs, investors, because they are the most likely to be able to interpret what it is for, namely, the successful introduction into the economic process of technological, commercial, or organizational improvements" (CWL 15, 133, note 186). See also Quinn's essay "Economics in the Anthropocene: Blue, Green, and Other Colors" and his reaction to Toni Ruuska's essay "Capitalism and the absolute contradiction in the Anthropocene" in chapter III.

analysis of economics issues. I was going to type, "but this is another story," but it is not. It is part of the mess we are in. Like the well-intentioned advisors to Pope Francis, Attenborough has no way of seeing the hole in the "doughnut economics" of Kate Raworth which he endorses. Add as many rings to your model as you would like, name as many minimum requirements as you would like, name as many ceilings you would like,[41] the model remains a model.

I admire Attenborough for speaking about the future. I believe there are things worth creatively recycling in what he writes about "rewilding," and I am not making him out to be a fool. His vision of rewilding the land and seas is optimistic and hopeful: "We can yet make amends, manage our impact, change the direction of our development and once again become a species in harmony with nature. All we require is the will."[42]

Well, the last six words are simply not true. In addition to good will, we require understanding as well as understanding of understanding—which would include, for example, understanding the biochemistry of what we are doing when whatting, and the biochemistry of intending adventure when whatting about the future, asking about what might be.

Since I am not Damiel or Cassiel,[43] my little insights—what we call in Spanish "*chispazos*"—are into, and in some important ways, "in" the phantasm,[44] which, like wonder, is a fascinating blend of nerves, molecules, and patterns of physics that I can name (as I just did) without understanding more than how to use the words *nerves*, *molecules*, and *patterns of physics* intelligently.[45] The fact that wonder is organic is not a suspicion; it is phenomenologically verifiable. The fact that

---

[41] "This new ring holds the minimum requirements of human well-being: good housing, healthcare, clean water, safe food, access to energy, good education, an income, a political voice and justice. It hence becomes a compass with two sets of boundaries. The outer ring is an ecological ceiling below which we must remain if we are to have a chance of maintaining a stable and safe planet. The inner ring is a social foundation that we must aim to raise everyone above to enable a fair and just world." *A Life on Our Planet*, 128.

[42] A Life on Our Planet, 220.

[43] Damiel (Bruno Ganz) and Cassiel (Otto Sander) are two angels that are "outside" human space and time in the film *Wings of Desire* (1987).

[44] "The act of insight is into the sensible or the imagined." Aristotle, *De anima*, III, 7, 431a 14–16; 8, 432a 3–10. See also "Insight into Phantasm" in Bernard Lonergan, *Verbum: Word and Idea in Aquinas*, ed. Frederick Crowe and Robert Doran, Collected Works of Bernard Lonergan 2 (Toronto: University of Toronto Press, 1992), 38–46.

[45] See further "Nominal and Explanatory Definition," CWL 3, 35–36.

humans wonder about images is a no-brainer for kindergarten and primary teachers. Remove crayons, white boards, CD players, movement, scents, drawings, tales, pictures, and other images from the classroom, and children will have a hard time learning. So, yes, significant reforms in education are needed to help Rita get the rhyme wrong.[46]

Sane education is a fitting name for raising the question marks of children so that one fine day they are effective leaders—characters cajoling, encouraging, and persuading others to take a stand in favor of educating Rita to get the rhyme wrong and Rita right, a stand in defense of the little one's sacred, mighty "mightn't it?"[47] So much depends on what a baby is and how he or she or we might "late in life, with indomitable courage, continue to say that we are going to do what we have not yet done: we are going to build a house."[48]

The house of researchers, interpreters, historians, dialecticians, founders, policy-makers, systems-planners, and communicators is to be, I believe, a house of wondrously, marvelously mediated meanings, including feedback and "feedforward" from the masses.[49] The concern for managing meaning articulated by the authors of "The Anthropocene: Comparing Its Meaning" is timely and

---

[46] I am remembering a scene from the film *Educating Rita*. Frank (Michael Cane). "Do you know Yeats?" Rita (Julie Walters): "The wine lodge?" Frank: "No, W.B. Yeats, the poet." Rita: "No." Frank: "Well, in his poem 'The Wild Swans at Coole,' Yeats rhymes the word 'swan' with the word 'stone.' You see? That's an example of assonance." Rita: "Yeah, means getting the rhyme wrong."

[47] "Ursula (three years, five months) is continually offering absurd suggestions about various things. 'That *might* be for so-and-so, mightn't it?' and when her mother says, 'Yes, but—' she always persists, 'But it might, mightn't it?'" Susan Isaacs, *Intellectual Growth in Young Children* (London: Routledge & Kegan Paul, 1930), 360.

[48] Gaston Bachelard, *The Poetics of Space*, Boston, Beacon Press, 1969, p. 61.

[49] The $C_{i,j}$ diagram on page 120 of "Structuring the Reach Towards the Future" intimates the spread of mediations. Note that the symbol C_9 in that diagram and in the staircase diagram compactly includes every brand of common sense, the colorful, multi-tongued concrete plurality of approximately 7.8 billion humans, "the almost endlessly varied sensibilities, mentalities, interests, and tastes of [hu]mankind" (CWL 14, 135). Popular feedback is symbolized $C_{9,8}$, while popular 'feedforward' is symbolized $C_{9,1}$. The "problem of the creative use of the available media," which now includes social media, and "the task of finding the appropriate approach and procedure to convey the message to people of different classes and cultures" (CWL 14, 135), requires backup, for "without the first seven stages there is no fruit to be borne" (CWL 14, 327). The first seven stages are conversations symbolized $C_{i,i}$ and $C_{i,i+1}$ (i = 1 to 7).

gives rise to a set of related questions. What is meaning? Are there discernible different carriers, functions, realms, even stages of meaning?[50] What am I doing when me-ning, with or without typed or signed words?

What happened to Helen Keller when she leapt in and from years of mooded and sometimes moody experiences of bubbles and baths, tasting and swallowing water, to signing the word *w-a-t-e-r*?[51] What was going on with Henry Cavendish in his experiments on the formation of water from "inflammable" air (H_2) and "dephlogisticated" air (O_2)? What was going on with Antoine Lavoisier when he added further clarity to what Cavendish had discovered? What goes on in and to you or me when we what or why about a sequence of visual, auditory, gustatory, olfactory, or tactile images? Oh, what's going on? Yes, yes, yes, yes, yes.

Gaseous and Intelligible Emanations

An image that I find useful to imagine the possibility of rewilding human dynamics is the phenomenon of a baby laughing. When a baby laughs, it does so from toes to head, laughing with butt muscles and all other muscles. Legs kick in the air, hands wave, and there might even be a release of bodily fluids and/or a gaseous emanation. It is quite a phenomenon that reminds me of a quote from Picasso about painting integrally. At an early age he "could paint like Raphael, but it has taken me a lifetime to learn how to paint like a child." The fantasy of recovering baby *anthropos* in the decades and centuries ahead pivots on serious, sexual, dreamy understanding,[52] that which leads me to skip meals, lose track of time, and wake from dreamworld, sometimes at an early hour, with a relevant image that just might resolve a theoretical or existential problem.

---

[50] For an introduction, see chapter 3 "Meaning" in Bernard Lonergan, *Method in Theology*, ed. Robert Doran and John Dadosky, Collected Works of Bernard Lonergan 14 (Toronto: University of Toronto Press, 2017), 55–95.

[51] See McShane, *A Brief History of Tongue*, 31–36. See also note 56 below.

[52] "If my intelligence is mine, so is my sexuality. If my reasonableness is mine, so are my dreams." CWL 3, 499. Part of the rewilding task is to expose and transform a decadent Judeo-Christian ethics of sexuality in our interpersonal meetings and greetings. "Does my sexiness upset you? / Does it come as a surprise / That I dance like I've got diamonds / At the meeting of my thighs?" (Maya Angelous, "Still I Rise") Healed, restored, non-offensive, sexy "good will wills the order of the universe and so it wills with that order's dynamic joy and zeal." CWL 3, 722. See also Philip McShane "Appendix: Rescuing Sexuality," in the epilogue of Patrick Brown and James Duffy (eds.), *Seeding Global Collaboration* (Vancouver: Axial Publishing, 2016), 241–45. The question I pose at note 68 below regarding the terminal of human history is relevant.

Part of the negative Anthropocene mess we are in is the common view that concepts precede understanding and can be integrated in "conceptual maps" (or "mental maps"). In those contexts, *concepts* are simply names, for example the names of various disciplines. We are not quite sure when or how these concepts emerged, either historically or personally, but we use them, analyze them, classify them, and connect them using arrows or other figures.[53]

There is, however, a distinct meaning of *concept*, not something I memorize, but something I become, after embracing and living for days if not weeks or months questions that occur to me, for example, "What is time?" "What is capital?" or "What is happening when I ask what?"

Now the questions, as well as the little insights along the way, are conscious, so more like indigestion than digestion. The concept proceeds from an *intelligible emanation*,[54] an "inner process" that happens, for example, when I hear and understand a joke, solve a puzzle, or figure out "who done it." An intelligible emanation is not a gaseous one, although the two might coincide if the release to the tension I experience when understanding coincides with an abrupt change in CO_2, H_2, or N_2 levels. It is possible to expel gas while laughing or learning integral calculus. What's more, the release of tension is phenomenologically verifiable, for instance when I hear a joke and get it. Likewise, the process leading up to an explanatory concept is also phenomenologically verifiable and memorable.

> There are two characteristics of a serious explanatory concept. You will remember the weeks, months, even years, which you spent—with feats of curiosity, not feats of memory—in struggling towards it. You will be able, even years later, to speak of it coherently, illuminatingly,

---

[53] For example, like figure 3 in Zalasiewicz et al., "The Anthropocene: Comparing Its Meaning," as well as figure 3 in "The emergence and evolution of Earth System Science."

[54] This is a translation of the Latin emanatio intelligibilis: Non ergo accipienda est processio secundum quod est in corporalibus, vel per motum localem, vel per actionem alicuius causae in exteriorem effectum, ut calor a calefaciente in calefactum; sed secundum emanationem intelligibilem, utpote verbi intelligibilis a dicente, quod manet in ipso. Thomas Aquinas, Summa theologiae, I, q. 27, a. 1 c: "Procession, therefore, is not to be understood from what it is in bodies, either according to local movement, or by way of a cause proceeding forth to its exterior effect, as, for instance, like heat from the agent to the thing made hot. Rather it is to be understood by way of an intelligible emanation, for example, of the intelligible word which proceeds from the speaker, yet remains in him." See also "Emanatio Intelligibilis" in Lonergan, Verbum, CWL 2, 46–59.

through illustrations, for perhaps ten hours. Maybe you are led by this to suspect that serious explanatory concepts are rare achievements? And certainly they are not passed on from generation to generation in compact little learned nuggets.[55]

The desire to understand is spontaneous, evident in the questioning poise of children. "How?" "Why?" "What?" and "What might be?" do not need to be taught. Young children can and do ask questions about little flowers and little princes, also about the big picture, the story of everything, all-stories, where "all" includes the child's natural quest: "Mommy, what is *love?*" "Daddy, what does *mean* mean?" Sadly, sometimes precious questions are brushed aside.

> We adults usually turn aside a child's challenge with an irritated "Oh, you know what I mean!" How intimidating, how unfair, how desensitizing that response of annoyance can be! If we ever stopped to reflect seriously and honestly, it might become clear to us that, often enough, there really wasn't anything clear that we could be said to have meant.[56]

And what do I have to say clearly about time,[57] not to mention organic development? What do I have to say about how this baby might be educated to become, late in life, an elder, a wise guy or gal capable of discerning the implicit discernings in those involved in collecting data (research), interpreting data, and creating narratives, thus presenting a fantastic version of the past 300,000 years,

---

[55] McShane, *Economics for Everyone* (3rd edition), 22.

[56] Gareth B. Matthews, *Philosophy and the Young Child* (Cambridge, MA: Harvard University Press, 1980), 21. At their best, mediators of meaning do their best to remember, self-taste and self-possess, their early breakthrough to signing *hi* or *bye-bye*, *mama* or *dada*, or *juice* or *milk* with fingers or tongue. Remembering such a passage from moody confusion and frustration to making intelligent noises helps one to distinguish child-friendly, child-out education from techniques of parroting to pass an exam, making "conceptual maps," and other truncated techniques that are popular nowadays, now-a-daze.

[57] It was not that long ago that some bright thinkers figured out that the "absolute time" and "absolute space" of classical mechanics are not absolute, thus contributing to a problem confronting the curious, extroverted animal who uses the word "is" countless times each day with disassociating yea-saying and naysaying from pointing at what is "already out there now" in front of my eyes. "To be" does not mean "to be within space and time," and "interpretations of being in terms of space and time are mere intrusions of imagination" (CWL 3, 403–404). The conundrum was expressed by Eddington identifying two distinct tables—one solid and colored on which he worked, the other a manifold of "wavicles" in mostly empty space.

something better than what was going on?[58] Obviously, or sadly not, asking about the intelligibility of space and time is much simpler than fantasizing wise elders.

Concrete Fantasy

Concrete fantasy is tough though vitally important work,[59] and I do not believe it does those of us concerned about the Earth's future any good to delegate the work to science fiction writers. The transition to a positive Anthropos-scene entails a recovery of baby Jane and baby James, perhaps the strangest "things"[60] in the glorious cosmos. "Understanding these human dynamics is essential for the effective guidance systems required for steering the future trajectory of the system."[61] Concern for rigor and clarity is not to be tossed out the window, but there needs to be a growing appreciation that logic does not rule the roost, that creative recycling is needed to listen and speak to the nearly endless variety of sensibilities, interests, ages, mindsets, and temperaments that walk the globe.

---

[58] There is a dense paragraph about a possible "geohistorical ethos of discerning discernings of discernings" in McShane's essay "Crecycling *Sustainability*" on page 93. See also note 69 below. Collaborative science fiction will reach something of a high point in the tasks labeled D_i and F in the staircase diagram. Disciples of Lonergan tend to confuse dialectic (D_i) with an encounter with thinkers who are dead, and/or they avoid encounter altogether. See Patrick D. Brown, "Functional Collaboration and the Development of *Method in Theology*, Page 250," *Himig Ugnayan: A Theological Journal of the Institute of Formation and Religious Studies* XVI (2015/2016), 171–99.

[59] "Without fantasy, all philosophic knowledge remains in the grip of the present or the past and severed from the future, which is the only link between philosophy and the real history of mankind." Herbert Marcuse, *Negations: Essays in Critical Theory*, translated by Jeremy J. Shapiro, Boston, 1968, 155.

[60] See "Things," CWL 3, 270–295. See also Meghan Allerton and Terrance Quinn, "The Notion of a Thing," *Journal of Macrodynamic Analysis* 14 (2020), 95–109.

[61] "The big challenge is to fully integrate human dynamics, as embodied in the social sciences and humanities, with biophysical dynamics to build a truly unified ESS effort. … Although long-ignored by the physical perspectives that have dominated ESS, understanding these human dynamics is essential for the effective guidance systems required for steering the future trajectory of the system." Steffen et al., "The Emergence and Evolution of Earth System Science," 61. In a chapter titled "A Heady Folly," McShane draws a witty and not altogether far-fetched analogy between Heady Lamar's invention of a torpedo-guidance system in the 1940s that depended on what Lamar called "frequency hopping" and Lonergan's discovery of what McShane calls "functional hopping" in 1965. *The Allure of the Compelling Genius of History: Teaching Young Humans Humanity and Hope* (Vancouver: Axial Publishing, 2015), 53–63.

To arrive in a never-never land somewhere over the axial rainbow, we little humans need to love and live a trinity of questions as if life before death depended on it: "When did I last truly, deeply, madly understand? When did I last truly, deeply, madly speak? When did I last truly, deeply, madly listen?" The vitally important, *anthropos*-caring, cosmos-hugging action is to recover a time of questions.

When the child was a child,
it was the time of these questions:
Why am I me, and why not you?
Why am I here, and why not there?
When did time begin,
and where does space end?
Isn't life under the sun
just a dream?
Isn't what I see, hear and smell just a
mirage of the world before the world?
Does evil actually exist,
and people who are really evil?
How can it be that I, the one I am,
wasn't there before I was there
and that sometime I, the one I am,
no longer will be the one I am?
The child needs oxygen.
Breathe deep down.[62]

While there is a growing body of literature on the "post-Anthropocene,"[63] throwing baby *anthropos* out with the dirty bathwater would be suicide. "The child needs oxygen," and the predicament we are in challenges kindergarten, primary, and secondary teachers as much as it challenges the rest of us. The origin of insights, as well as new words and new expressions, is every child's wonderbone. Little ones are at home in wonderful, wonder-filled 'boom chicka boom' song and dance wordplay. They give their pencils a twist and letters spread out, become a

---

[62] *Wings of Desire* (1987). This film begins with a fairy-tale-like narration of a time when children were children and asked important questions. Damiel and Cassiel cannot participate in the unfolding events; they only pretend. Children participate and pretend.

[63] See note 21 in the Preface.

fish or a bird and fly away.[64] Twisting and turning pens, colored pencils, and crayons is what four- and five-year-olds do. Twisting and turning (educating) children to spread out letters to become brooks and birds, possibly even words talking about themselves, is not a new-fangled "student centered" technique; it is a dream about little birds growing tails, ruffling out feathers, and flying off.

The challenge of recovering human dynamics—a revolutionary core of cosmic process—implies a recovery of baby laughter, both ontically and phyletically, so slowly, one baby step at a time, even though the "times" are not exactly on our side: "Modernity, and the economics and political [and educational] ideas at its core, are premised on a separation of humanity from nature."[65] It is time to go back to the drawing board to ask foundational questions.[66]

> *When I was young, it seemed that life was so wonderful*
> *A miracle, oh it was beautiful, magical*
> *And all the birds in the trees, well they'd be singing so happily*
> *Oh joyfully, playfully watching me*[67]

What might it mean to understand human history in all its richness and variety, if the terminal of history, her-story, all-stories is cosmic joy and zeal,

---

[64] "Often as I write some Greek letter, Theta or Omega, I have only to give my pen a twist, and the letter spreads out, to become a fish, and I, in an instant, am set thinking of all the streams and rivers in the world, of all that is wet and cold; of Homer's Sea, and the waters on which Peter walked to Christ. Or else the letter becomes a bird, grows a tail, ruffles out his feathers, and flies off." Herman Hess, *Narcisuss and Goldmund* (New York: Farrar, Straus and Giroux, 1968), 62.

[65] *A Multidisciplinary Approach*, 168. "And what is modern about modern mind, one may ask, if Hegel, Comte, or Marx, in order to create an image of history that will support their ideological imperialism, still use the same techniques for distorting the reality of history as their Sumerian predecessors?" Eric Voegelin, *The Ecumenic Age*, Louisiana State University Press, 1974, p. 68.

[66] "As with other new historical frameworks (labor history, gender history, and structuralism, for instance), Anthropocene history has sent historians back to the drawing board, asking foundational questions about the purpose of history, the appropriate forms of evidence, and how meaningful stories might be crafted out of reality's vast messiness." Thomas, Williams, and Zalasiewicz, *A Multidisciplinary Approach*, 127.

[67] Supertramp, *The Logical Song*, Breakfast in America (The Village Recorder, 1979).

children young and old watching, singing, dancing, playfully, joyfully?[68] As McShane noted in "Ant Hop," cherishing what's what is the core opener to the positive Anthropocene.

The challenge before us is to figure out what steps we might take now to increase the probability of a better life on planet Earth for our descendants in the year 3522,[69] even though the window of opportunity for mending our ways might be 30 or less years. This is very difficult, for it strains molecules and challenges inertia that has accumulated for 200 or 300 years, if not longer. Our professional identities and sabbatical plans are being challenged by a glocal predicament[70] calling forth glocal collaboration and psychic adaptation that is analogous to starling murmuration. To fly thusly, we baby humans need to "adapt speed and velocity"[71] and psychically adapt to one another as we solve large problems by breaking them down into little ones and as we encourage, both in one's nearest neighbor (oneself) and in one's students and colleagues, the patience to live one's questions and find out, personally and slowly, both what is happening when I what in any area of my life, and how an accumulation of such happenings might lead to an increment of progress. Such adaptation evokes Socratic humility[72] about what

---

[68] I have significantly altered this question: "What does it mean to understand human history in all its richness and variety, if the terminal of history is the Anthropocene?" Thomas, Williams, and Zalasiewicz, *A Multidisciplinary Approach*, 131.

[69] Goethe remarked: "He who cannot draw on three thousand years is living from hand to mouth." Half of 3,000 is 1,500, and 2022 + 1500 = 3522.

[70] "The new existential challenge is global, but many of our attempts to understand and cope with it will be local." *A Multidisciplinary Approach*, 197. Taking responsibility for the indeterminate future of the Earth might entail slowing down to attempt "apparently trifling problems" (CWL 3, 27), changing sabbatical plans, perhaps even canceling some plans. In the spring of 2019, the organizers of "The Positive Anthropocene" conference scheduled for July 2019 (see Appendix B) decided that conserving resources outweighed the possibility that the event would be an effective and beautiful intervention in historical process and decided that it was best to cancel the conference.

[71] See note 25 in the Preface.

[72] Arne Næss writes this about arrogant stewardship: "The arrogance of stewardship [as found in the Bible] consists in the idea of superiority which underlies the thought that we exist to watch over nature like a highly respected middleman between the Creator and Creation" (*Ecology, Community and Lifestyle: Outline of an Ecosophy* [Cambridge University Press, 1989], 187.) His hermeneutics of suspicion is not balanced by a hermeneutics of recovery of his baby-best self, raising and answering questions. (See note 24 in the Preface on the two hermeneutics.) He does not identify the possibility of

sewn in the cosmos. The global predicament requires "crecycling your own position regarding the state humanity is in."[73] In good time, a strange redemption of time will be made possible by a glocal flock of luminous whatters creatively remembering the future better than it was.

---

humble stewardship that would help him find in the book of Genesis intimations of what and what might having been sewn in the cosmos.

[73] McShane, "The Masses and Sustainability," 111.

Appendix A: Abstract for "Structuring the Reach Towards the Future"

Philip McShane

January 29, 2019

My efforts are in line with pointers towards such a structure given by Arne Næss, Bernard Lonergan and Peter Drucker in the last century. The heuristic is of a double structure of collaboration. The zones, to be homed in on by individuals, according to their background, are: 1. research, 2. interpretation, 3. verified story, 4. dialectic assessment, 5. foundational dynamics, 6. policy structuring, 7. systematic foresight, 8. effective communications. These obviously require spelling out in their cyclic dynamic, but the interest of the Helsinki gathering is evidently future bent, so my presentation would be foundational, looking at a realism of the collaborative efforts of 6, 7, and 8. The aim is the integration of global groupings towards political effectiveness.

What is important for me now is to give some glimpse of the effective dynamic, and it seems best to do so by presenting first two diagrams that stir the imagination towards creativity and optimism. I present them here on the second page. The first diagram present's the brilliant achievement of Archimedes: a way of screwing up water.

The issue is to take seriously Næss's and Lonergan's appeals for a collaboration of a quite precise scientific structure, indeed a cyclic structure described by Lonergan back in the mid 1930's.

> But what is progress? It is a matter of intellect. Intellect is understanding of sensible data. It is the guiding form, statistically effective, of human action transforming the sensible data of life. Finally, it is a fresh intellectual synthesis understanding the new situation created by the old intellectual form and providing a statistically effective form for the next cycle of human action that will bring forth in reality the incompleteness of the later act of intellect by setting it new problems.[1]

Western intellect emerged in a split fashion, due considerably to the influence of Aristotle's view of science which grounded a view of science as the verification

---

[1] "Essay in Fundamental Sociology", in *Lonergan's Early Economic Research*, edited with commentary by Michael Shute, University of Toronto Press, 2010, 20.

159

of theory in data, where the ideal of theory was an axiomatics. Primitive science was a pragmatic affair, measured by success. A fishing weir is a bright idea: but will it better provide more fish? Modern business brought forth this poise, and perhaps we might think of Peter Drucker as lifting it towards the level of science: think, thus, of the three realms of management: Policy, Planning, Executive strategies. Now pause over the diagrams.

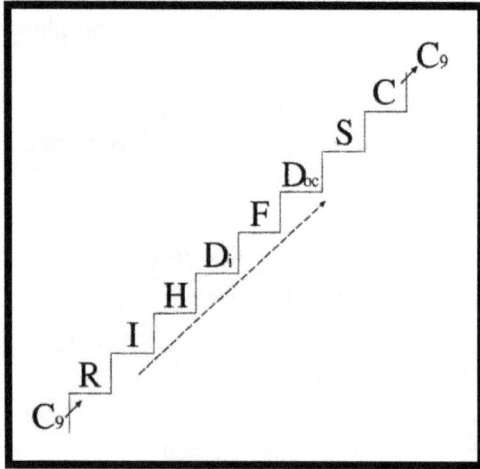

The structure is a ten-step collaborative identification which would locate all our present efforts towards what I call *The Positive Anthropocene* in dynamic cohesion. I think of Adam Smith and the making of pins, but now the issue is effective pens, in both senses of that word. There is to be a structured ink-articulateness and also *pens* in the sense of layered operative unities. Perhaps this brings to mind the venture of communism with its cells, and rightly so, for that is an inspiration for the effort of local divisions. My lengthier discussion of the matter may well be called *A Common Quest Manifesto*.[2]

Space does not allow elaboration on [1] adding global community involvement, C_9, [2] "filling the gap" between the first three cyclic components and the final three. These are the two zones of 4. Dialectic Assessment and 5. Foundational Creativity which link Aristotle's three with Drucker's, fill out on missing elements in Næss, and fulfill Lonergan's idea of a "Fundamental Sociology."

---

[2] *Æcornomics* 3: A Common Quest Manifesto, available at:
http://www.philipmcshane.org/ecornomics.

Appendix B: The Positive Anthropocene Conference

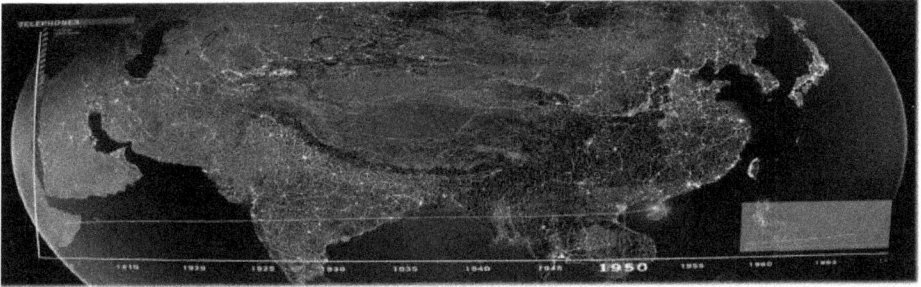

The Positive Anthropocene

University British Columbia   July 8-12, 2019

Conference Topics

- Implementation
- Global Leadership
- Economic Reorientation
- Detruncating Education
- Initiating the Spiral of Progress

Fee and Accommodations

- Conference fee: $75 CDN. Includes break provisions, reception and final dinner, and a copy of *Divyadaan* 30/1 (2019) "Religious Faith Seeding the Positive Anthropocene Age"
- Carey College housing : http://live.carey-edu.ca/rooms
- UBC housing :information@housing.ubc.ca
- UBC Pacific Spirit Hostel : http://www.suitesatubc.com/
- TransLink Vancouver transport : https://www.translink.ca/Fares-and-Passes/Fare-Pricing.aspx

"The Anthropocene Age, recently formally recognized, is the period of evolution in which humanity's discoveries and decisions give manifest tone to evolution's dynamic. The early negative Anthropocene was a stumbling period of local tribalism without large-scale warping of global dynamics. The following axial period added not only large-scale toxicities to evolution but also a solid and growing detriment to humanity's neurocraving in twisted grammar-structurings and truncated educational identifications of the dynamics of humanity. You and I need a Dionysian fantasy to begin to seed the distant positive Anthropocene."

Philip McShane (M.Sc., Lic. Phil., S.T.L., D.Phil [Oxon]), director of the conference.

Further information & registration:
robert.henman@msvu.ca

Sponsored by
St. Marks College

The Positive Anthropocene Age:
Seeding a New Popular Culture

International Conference July 8–12, 2019
Corpus Christi College, Vancouver, British Columbia

My *Webster's Dictionary* gives as first meaning of conference, "the act of consulting or conversing on a serious matter," and so shall this gathering be. Prior to the conference, people may certainly exchange papers or drafts of papers presenting their stands or their questions on the topics broadly raised for each of the five days.

The serious matter is what is named in the title. The serious matter will be taken in five stages, detailed better as we move along. Participation need not involve attendance at the gathering: conferring at a distance is welcome all the way to next July and beyond.

Let me first briefly name our five problem-topics: (1) implementation; (2) global leadership; (3) economic reorientation; (4) detruncating education; and (5) initiating the spiral of progress. Next, I express my own initial take on these. I would note that the reach, from now until long after the conferring, is for concrete cosmic guidance towards seeding progress. Such concrete guidance is to blossom from scribbles of those willing to participate in the search symbolized by the smaller gathering in Vancouver.

Monday, July 8th
Our first-day focus is the effective transformation of "Common Meaning and Ontology" (CMO). The focus is to be mediated by some grip on Lonergan's "Standard Model" for such an inquiry: a commonsense grip on *Insight* (I) and *Method in Theology* (M). Our focusing, of course, aims to be effective in lifting our communal intentions with regarding to that grip. The day is to home in effectively on Lonergan's outrageous demands that metaphysics include "implementation" (I. 416) and that theology aims at "fruit to be borne" (M. 355).

Tuesday, July 9th
We shift to the crisis of religious leadership in that effectiveness, perhaps neatly posed by Lonergan in *Phenomenology and Logic*, first, the positive challenge: a leadership "to be a resolute and effective intervention in this historical process" (*Phenomenology and Logic*, 306); secondly, the failure of a leadership, assumed to be tuned to the Field of Being, but who operatively "retire into an ivory tower and

exert no influence upon society at large: then we have a situation where the people who can do the most harm are doing it and the people who could do the most good are not" (306–7). The recent history of this embeddedness in the second stage of meaning is the topic of the first four articles of *Divyadaan* 30 (2019), which is to be supplied.

Wednesday, July 10th
The issue concerning us on this third day is indicated by the title of the fifth article in the *Divyadaan* volume just named: "Finding an Effective Economist: A Central Theological Challenge." It will be 51 years in 2019 since Lonergan made the request, "find an economist." This is perhaps the most obvious instance of failure to implement, to bear fruit. So, it will be the most revealing of the defective procedures of our times. On page 766 of *Insight*, Lonergan wrote that "theology possesses a twofold relevance," but the relevance is at present vague, ineffective, "effete" (M. 99). Our challenge is to grasp effectively that ineffectiveness in ourselves and in our cultures. The accumulation of a fresh existential poise in these three days is to bring us to concrete strategies of the necessary finding of economists. Obviously, the strategy includes enlightening conversations in the face of present blockages.

Thursday, July 11th
So, our climb in understanding our present standing and the remedying of it slowly and globally, leads us deeper. The failure noted and ingested on the third day is certainly rooted in the difficulty of educating economists of our acquaintance. What is this failure in its fullness and how might we lift culture to a new center of gravity, of seriousness? "That shift in the center of gravity, that habituation to a differentiated consciousness, is the fruit of education" (*Topics in Education*, 116). The issue is to search for and find—cyclically—the recurrence-schemes of ontic and phyletic shifting.

Friday, July 12th
The cyclic issue was nursed as a problem by Lonergan since his 29th year: the problem of "providing a statistically effective form for the next cycle of human action that will bring forth in reality the incompleteness of the later act of intellect by setting it new problems." (*Essay in Fundamental Sociology*, 1934). So, we return to the first day's topic, but with the "cumulative and progress results" (M. 4) of a communal "invitation to understand something about the process of human history, and a summons to decisiveness at a rather critical moment in the

historical process." (*Phenomenology and Logic*, 300). We return to the fifth chapter of *Method*.

*****

Further details of the conference dynamics will thus emerge. But it seems neat to conclude with my own basic suggestion: that the focus of the conference should be on implementation, on FS_8, but bracketed by a foundational perspective on it (so, a personal triple-positioning on it—my concluding paragraph will capitalize *it*—à la *Lonergan's 1833 Overture*) and by constructive functional attention to C_9. That constructive attention is to be massively creative. The articles in *Divyadaan* 30/1 (2019) drive us towards thinking out redemptive isomorphic structures for contemporary situations and topologies of psychosocial structures and their referents.

But now, during "the animalization of man on the higher level of his achievement" (*Analytic Concept of History*, 1934, conclusion), we must face, with graceful gallantry and gantry, towards a genesis of quite novel pyromorphic structures. "It is, I fear, in Vico's phrase, a *scienza nuova*" (*Shorter Papers*, 223). It? IT? Has not this IT something to do with the IT of page 232 of *The Allure of the Compelling Genius of History*? Is IT not to weave forward the website series, "Questions and Answers" (http://www.philipmcshane.org/questions-and-answers) with concrete effective **I**nternal **T**opologies that would fire up a global meaning, for example, of Question 51, "You Make My Skin Caul" and Question 56, "Breaking Forward to Global Care"? So that our conferring might lean us, seedily, globally, towards an Effective Interior Lighthouse Poise of "IT Makes Our Skins Caul"?

Phil McShane
November 2018

Index

#AGoodAnthropocene, 76–78

Aesop, 75

Anthropocene
  as a geological unit, 139, 143, 145
  epoch, 43, 60, 62, 66, 73, 85, 145
  negative, vii, 4, 13–15, 26–30, 63–65, 71, 76, 79, 94, 104–105, 127, 147, 151
  positive, vii–viii, 4, 26, 31, 34, 42, 63–66, 71–72, 79, 82, 94, 104, 135, 140, 153, 156, 160
  starting date, ii, 4, 12, 22, 135
  the term, ii, vi, vii, 3, 9, 11, 14, 23, 60, 71, 139, 142–145

Anthropocene Project, 43, 76, 77

Anthropocene Working Group (AWG), ii, 11, 139, 142–145

Archimedes' screw, 30–31, 72–73, 95, 102, 105, 115–118, 159

Aristotle, 3, 13, 116–117, 129, 135, 146, 148

Attenborough, D., iii, 147–148

axial
  humanity, 122, 129
  period, 4, 104, 134–135

Babel, I., 82

Baichwal, J., iii, 43, 76–78

Baumgart, B., 65

Beckett, S., 94, 132

bias
  general, 103–104
  group, 103

bionoosphere, 13, 15

biosphere, i, 13–15, 25

Blair, E., 65

Bonnedhal, K. J., 29

Boulanger, N., 102

Brown, D., 79, 80–82

Burtynsky, E., iii, 43, 76–79

capital, capitalism, 22, 26, 43, 46–51, 57, 124–133

chronostratigraphy, -graphic, 142, 143

climate change, i, 21, 23, 46, 77, 141

collaborate, -tion, iv, v, 42, 47, 57, 96, 111, 120, 127, 131, 142, 145–146, 156, 159, 172, See also division of labor

common sense, 37, 97–99, 101, 103, 130, 149

concept, -ual, 89
  Anthropocene as, 61, 66
  constructs, 60
  maps, 151
  tightly defined, 143
  two opposed meanings of, 151

cosmopolis, 105, 133

crecycling, 8, 30, 31, 47–48, 89–93, 98, 111

Crutzen, P., ii, 140, 143

de Pencier, N., iii, 43, 76, 77–78

deep ecology, 57, 133

Deep Green Resistance, 25, 55, 56, 93

development
  economic, 41, 62
  human, 42, 101, 102, 126
  of societies and cultures. See progress
  organic, 126, 152
  sustainable, v, 64, 76
  technological, 15
  urban, 12

dialectic, dialecticians, 93, 102, 104, 121, 123, 125, 130, 157, 166

division of labour, 15, 30, 47, 57–58, 89–93, 106, 127, 129, 132

Drucker, P., 116, 121–122, 131, 159–161

Earth System, 3–4, 11, 12, 61

Earth System Science, -tists, i, vii, 23, 142–144, 151, 153

economic, -s, 50, 91
  activity, 51, 66
  actual, 42, 50, 51, 64, 112
  analysis, 61, 148
  circular, 147
  doughnut, 148
  ecological, 23, 41–42
  global, 41, 45–46, 49, 52
  growth, 62, 64
  reorientation, 164
  sustainable, 41, 64
Economics for the Anthropocene, E4A, 41–42
economist, -s, 23, 61, 165
ecosystems, i, 22, 25, 41, 43
education, 66, 72, 79, 100, 149, 154–155, 165
  capitalist, 131
  failure in, 76, 99, 101, 129
  higher, 130, 133, 146
  history of, 98
  reforms in, 149, 164
Einstein, A., 30, 80
Eliot, G., 102
Farah, P. D., 34, 66, 128
feedback, 60, 76–79, 82, 99, 121
Fermat's Last Theorem, 91
foundations, foundationalizers, 43, 97, 98, 119, 121, 123, 127, 131
Fremaux, A., vi
Gaia, 30, 95, 122, 125, 136
Gandhi, 57, 89, 143
genetics, genetic method, 125–128
geological epoch, ii, vii, 23, 139
geological record. *See* stratigraphic record
geology, geologists, vii, 4, 9, 14, 57, 139, 143
Glikson, A.Y., vi
Global Boundary Stratotype and Section Point. *See* golden spike
golden spike, iii, vii, 139, 145, 146
Gore, A., 110

Gramsci, A., 119
Great Acceleration, 12, 66
Gross Domestic Product, 22, 41
growth. *See* development
Hegel, G. W. F., 129, 135
Heikkurinen, P., v, vi, 7, 8, 9
hermeneutics
  of recovery, vii, 133, 153, 155–156
  of suspicion, vii, 144–145, 156
Hesse, H., 136
heuristics, 35, 62–64, 83, 135
  genetic, 127
  of collaboration, 47, 120, 159
  of growth, 63–64
  ten-step diagram, 107, 110–111, 118
Higgs particle, 57, 92
historical process, 15, 56, 93, 119, 128, 155, 166
  intervention in, vi, 85, 95, 111, 124, 164
history, historians, 48, 90–94, 130, 144
Holocene, ii, 117, 135, 145, 147
Hopkins, G. M., 130
hyperobject, hyperobjects, 8, 56, 58
implementation, 66, 164, 166
Industrial Revolution, 7, 12, 50, 71, 130–131
International Chronostratigraphic Chart, 3, 11
International Commission on Stratigraphy (ICS), vii, 3, 11, 139
International Conferences
  The 3rd Peaceful Existence Colloquium, iv, 115, 117, 132, 139
  The 4th Peaceful Coexistence Conference, 141
  The Positive Anthropocene, iv, 85, 156, 163–66
International Union of Geological Sciences (IUGS), vii, 11
interpretation, interpreters, 48–52, 90–94, 121, 129, 132, 146
Jackson, M., 76

Jaspers, K., 4, 104, 134–135
Joyce, J., 75–76, 79, 82–83, 125, 136
Jung, C. G., 82
Keller, H., 97, 150
Kierkegaard, S., 8
Lawrence, J., 7, 51, 59–64
LeVasseur, T., iii, 7, 8, 55–57, 91–92, 118
Lonergan, B., vi, vii, 56, 90, 91, 92, 94, 96, 99, 101, 103, 106, 118, 129, 159, 161
Mann, G., 117
Marcuse, H., 122, 132
Marx, K., 121, 125, 128–135
mathematics, iii, 63
  as paradigm of science, 3
  storyform, 91
Maxwell, J. C., 92
meaning, -s, 8, 26, 49, 97
  common meaning, 164
  initial, 98, 104, 134
  managing, vii, 115, 142, 145, 149
  of *is*, 129
  of *meaning*, 97, 149, 152
  of *Renaissance*, 9
misothery, 8, 27, 55
Modernity, 144, 155
Mohorčich, J., 7, 25–27, 36
multidisciplinary, -arity, 142–144
Næss, A., 14, 57, 89–93, 96, 107, 119, 133, 156, 159
neoliberal, neoliberalism, 13, 51, 59–61, 112
Newton, I., 30, 92, 98–99
Nijinsky, V., 102
noosphere, 25–27
Ortega y Gasset, J., 14, 95, 109
Orwell, G., 65
Paris Agreement, 140
Phenomenology and Logic: The Boston College Lectures on Mathematical Logic and Existentialism, vi, 83, 85, 95, 101, 111, 145

physics, iii, 3, 30, 35, 47–48, 57, 91, 99, 122, 127, 129
Picasso, 150
Plato, Platonic, 95
Plutocene, 65
Pope Francis, iii, 140, 148
post-Anthropocene, vi, 154
Pound, E., 79, 85, 120, 126, 132
Praxisweltanschauung, 56, 57, 90, 93, 95
Priyadharshini, E., vi
progress, 7, 15, 47, 48, 64, 90, 106, 115, 127, 156, 159
  genetic control of, 127–128
  scientific. *See* science, statistically effective
Proust, M., 79, 124
Rantala, O., vi
Raworth, K., 148
reductionism, 98, 126
religion, religious, 8, 122,–125, 135, 143
research, researchers, 47–48, 90–94, 121, 146
Rilke, M. R., 36
Ruuska, T., vi, 7, 47–52, 60, 118, 121–134
Sale, K., 55, 57, 91–92
Sand, G., 132
Schrödinger, E., 92–93
science, 3, 65, 89–90, 117, 159
  Aristotelian, 3, 117, 159
  contemporary, 41, 42
  environmental, 43
  lower and upper, 104
  of history, 128–129, 134
  of the future, 30, 120
  primitive, 15, 117, 160
  pure, 13, 117, 123, 145–146
  statistically effective, 4, 30, 73, 91, 95–96, 106, 116, 127, 132, 159–160, 165
sequence, -ing
  of achievements, 91
  of collaborative communities, 132

of eight tasks, 121
of ideas, 63, 73, 90
of images, 150
of stories, 92
Shakespeare, 102, 120, 123–124, 131–133
Smith, A., 160
Snow, C.P., 144
socio-economic trends, 12, 22, 35
Socrates, 26, 36, 79
standard model, 92, 127
   economics, 41, 62
   of genetics, 127
Standard Model of particle physics, 91, 127
starling murmuration, vii, 156
Steffen, W., 22
stock markets
   algorithmic trading, 44
   global, 45
   high frequency trading, 45
   reports, 44
Stoermer, E., ii
story, storycheckers. *See* history, historians
storyform. *See* interpretation, interpreters

stratigraphic record, 4, 9, 11–15, 21, 139
Thunberg, G., iii
Toynbee, A., 4, 104, 135
Trump, D., 21, 27, 65, 104
Ulvila, M., 65
Valtonen, A., vi
Vico, G.
   scienza nuova, 82, 166
Voegelin, E., 4, 117, 128, 134–136
von Karajan, H., 75
watchers. *See* research, researchers
Waters, C., 11
what, whatting, 13, 30, 33–37, 41, 66, 71, 73, 77, 98, 101–102, 105, 135–136, 146–148, 151–154, 156
what's what, viii, 27, 72–73, 77–78, 82–83, 98, 134, 151, 156
Whitson, R. E., 136
Wilén, K., 65
Williams, M., 11
world religions
   convergence of, 85, 125, 136
Young, L., vi
Zakaria, F., 104
Zalasiewicz, J., ii, 11

About the Authors

Philip McShane (February 18, 1932–July 1, 2020) was an Irish mathematician, philosopher, economist, and theologian. He earned an M.Sc. in relativity theory and quantum mechanics with First Honors from University College, Dublin (1952–56), where he also lectured in mathematics. Later he did his D.Phil. at Oxford (1965–68). His dissertation was published as *Randomness, Statistics, and Emergence* (1970, 2nd edition, 2021).

McShane was a prolific scholar whose books focused on topics ranging from the foundations of mathematics, probability theory, and evolutionary process, to the philosophy of education and omnidisciplinary methodology. He also published introductory texts focusing on critical thinking, linguistics, and economics. In the area of methodology, he wrote *The Shaping of the Foundations* (1976), *Lack in the Beingstalk* (2006), and *Futurology Express* (2013); and in the area of economics *Pastkeynes Pastmodern Economics: A Fresh Pragmatism* (2000), *Piketty's Plight and the Global Future* (2014), and *Economics for Everyone: Das Jus Kapital* (1996; 2nd edition, 1998; 3rd edition, 2017). Among his introductory works are *Wealth of Self and Wealth of Nations* (1975, 2nd edition, 2021), *A Brief History of Tongue* (1998), and *Music That Is Soundless* (2005).

Recognized by many as the leading interpreter of the works of the Canadian philosopher theologian and economist Bernard Lonergan, McShane edited *For a New Political Economy* (Collected Works Bernard Lonergan volume 21, 1998) and *Phenomenology and Logic* (Collected Works Bernard Lonergan volume 18, 2001), and together with Pierrot Lambert he co-authored *Bernard Lonergan: His Life and Leading Ideas* (2013, 2nd printing).

James Duffy is an American Mexican scholar currently living in Morelia, Michoacán. After receiving a PhD in philosophy from Fordham University (1996), he went on to teach undergraduate philosophy and interdisciplinary studies at St. Mary's University of Minnesota. In the early 2000s, while on an extended sabbatical in Mexico, he was invited to participate in the design and implementation of a humanities and social sciences major at the *Instituto Tecnológico y de Estudios Superiores de Monterrey* (ITESM). He has also taught graduate students Pedagogy and Institutional Development at the *Universidad Nova Spania*.

In 2009, Duffy began an extended study of economic dynamics, and since then has published "Minding the Economy of Campo Real" (2018) and "*Fratelli Tutti* and Colorful Fruit to Be Borne" (2021). He also edited and wrote an introduction to *Religious Faith Seeding the Positive Anthropocene Age* (2019), a series of five essays written by Philip McShane.

Duffy has also published articles on foundations of probability theory, critical thinking, and the ethics of effective collaboration. In 2011, he co-organized and participated in "*Ampliando Nuestos Horizontes*" ("Expanding Our Horizons"), the First Latin American Lonergan Workshop in Puebla, Mexico. In 2014, he participated in the Sixth International Lonergan Conference, "Functional Collaboration in the Academy: Advancing Bernard Lonergan's Central Achievement," and together with Patrick Brown edited *Seeding Global Collaboration* (2016).

Currently Duffy is coordinating a series of ongoing dialectic exercises published in the *Journal of Macrodynamic Analysis*. These exercises bring scholars from different fields together to ask basic questions about intervening in the dialectic of history.

Robert Henman studied philosophy at Mount St. Vincent University and later went on to do his MA at the Atlantic School of Theology and did further graduate course work in the philosophy of education at the University of South Australia. He taught philosophy, Child Studies, and Peace and Conflict Studies at Mount St Vincent for 35 years.

From 1990 to 1994, Henman lectured on medical ethics at the Dalhousie Medical School in Halifax, Canada. From 2009 to 2014 he was the director of Lonergan conferences held alternately at the University of British Columbia in Vancouver and Saint Mary's University in Halifax. In 2018 he gave lectures on Lonergan's economics and education in Malaga, Spain.

Henman has published *The Child as Quest* (1984), *Global Collaboration: Neuroscience as Paradigmatic* (2016), and *Reorienting Education and the Social Sciences: Transitioning Towards the Positive Anthropocene* (2022, revised 2nd edition). He has also published articles in the areas of ethics, social science, psychology, meta-neuroscience, and peace studies.

Since his retirement from teaching in 2019, Henman has worked as an independent researcher and author, focusing on developing and communicating methods of intervention within the context of functional specialization. A complete list of his articles, books, and book reviews, as well as his CV, is available at http://roberthenman.com.

Terrance Quinn is a Canadian and American scholar currently living in Toronto, Canada. After receiving his PhD in mathematics, from Dalhousie University (1992), he did three years of post-doctoral work, one year at Trinity College Dublin, and then two years at University College Cork.

Quinn went on to faculty positions in mathematics at Texas A&M International (1995-2001), Ohio University Southern (2001-2006), and Middle Tennessee State University (MTSU) (2006-2018). At MTSU, he served as Professor and Chair (2006-2009) and Professor (2006-2018). In August of 2018, Quinn was awarded the rank of Professor Emeritus. He has held numerous positions of leadership in the academy, is active in publications, and serves on a number of editorial boards.

Many of Quinn's publications have been in mathematics, mathematical biology, and pedagogy in mathematics. He has two books on the foundations of science, *Invitation to Generalized Empirical Method in Philosophy and Science* (2017) and *The (Pre-) Dawning of Functional Specialization in Physics* (2017). In the last ten years, he has been branching into economics. With John Benton, he co-authored two books for general readership, *Economics Actually. Today and Tomorrow. Sustainable and Inclusive* (2019), and *Journeyism: A Handbook for Future Academics* (2022). Two papers (to appear, in 2023) are "On the operative presence of eight tasks in economics" and, in ecological economics, "An Implementable Methodological Solution to a Problem posed by Clive L. Spash and Others." A forthcoming book is *Advances in Heuristics of the Mechanical Structure of Two-circuit Economics*. The works of both Bernard Lonergan and Philip McShane have been influential. Currently, several projects in economics and education are in various stages of development. For more details, see Quinn's personal website: https://www.terrancequinn.com.

www.ingramcontent.com/pod-product-compliance
Lightning Source LLC
Chambersburg PA
CBHW060603210326
41519CB00014B/3556